COMPLEXITY ISSUES IN VLSI

Foundations of Computing

Michael R. Garey, series editor

Complexity Issues in VLSI: Optimal Layouts for the Shuffle-Exchange Graph and Other Networks, by Frank Thomson Leighton, 1983

COMPLEXITY ISSUES IN VLSI

OPTIMAL LAYOUTS FOR THE SHUFFLE - EXCHANGE

GRAPH AND OTHER NETWORKS

FRANK THOMSON LEIGHTON

The MIT Press
Cambridge, Massachusetts
London, England

Printed and bound in the United States of America

Library of Congress Cataloging in Publication Data
    Leighton, Frank Thomson
       Complexity Issues in VLSI
       (Foundations of Computing)
       Bibliography: p.
       Includes index.
       1. Electronic circuit design. 2. Microelectronics — Diagrams. 3. Graph Theory. I. Title.
    TK7867.L44 1983   621.381'73   83-7933   ISBN 0-262-12104-2

This book was prepared with the assistance of a National Science Foundation Graduate Fellowship, an MIT Bantrell Fellowship, Air Force contract #OSR-82-0326 and Defense Advanced Research Projects Agency contract #N00014-80-C-0622.

MIT Press

0262121042

LEIGHTON
COMPLEXITY VLSI

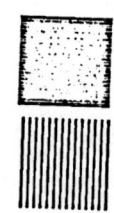

CONTENTS

FIGURES AND TABLES

SERIES FOREWORD

The foundations or theory area of computer science is concerned with the fundamental principles, concepts, and techniques underlying the computing field. Although at present this body of knowledge falls rather short of that possessed by more mature sciences like physics and chemistry, theoretical work in computing is expanding rapidly and much of value has already been learned. Progress in formal languages, computational complexity, and automata theory has had significant impact on the way a variety of computing problems are perceived and approached, and useful tools and techniques have followed from theoretical work. Even greater impact can be expected as succeeding generations of students carry the most recent products of the research community with them into industry.

By publishing comprehensive books and specialized monographs on the theoretical aspects of computer science, the series on Foundations of Computing will provide a forum in which important research topics can be presented in their entirety and placed in perspective for researchers, students, and practitioners alike. This inaugural volume, by F. T. Leighton, deals with the problems of laying out graphs or networks in the plane, as motivated by the design of circuits in Very Large Scale Integration (VLSI) technology. Although it represents work done as a doctoral thesis, it displays an unusual combination of insight into the heart of the problem, mathematical sophistication and inventiveness, and clarity of exposition. The volume includes techniques for deriving strong lower bounds on required layout area and maximum wire length and new layouts, which are both asymptotically optimal and practical, for the powerful shuffle-exchange networks. It is essential reading for anyone interested in learning about current state of the art in VLSI design theory.

Michael R. Garey

ACKNOWLEDGMENTS

The material in this book is drawn primarily from my doctoral thesis *Layouts for the Shuffle-Exchange Graph and Lower Bound Techniques for VLSI*. I submitted the thesis to the MIT Mathematics Department in September of 1981. The research was primarily funded by a graduate fellowship from the National Science Foundation. Additional support was provided by Air Force contract #OSR-82-0326, Defense Advanced Research Projects Agency contract #N00014-80-C-0622 and a postdoctoral fellowship from the Bantrell Foundation. I am deeply indebted to all four groups for their support during my tenure at MIT.

Many individuals contributed substantially, both professionally and personally, to the completion of this book. I am most grateful to Dan Kleitman, Gary Miller and Ron Rivest for serving on my thesis committee. All were of great help in preparing the thesis. Gary Miller, in particular, spent an extraordinary amount of time with me during my years as a graduate student at M.I.T. He was a great source of inspiration and encouragement, and was always available for discussion of my work. Sandeep Bhatt, Dan Hoey, Paris Kanellakis, Rao Kosaraju, Margaret Lepley, Charles Leiserson, Bob McKillip, Franco Preparata, John Reif, Michael Rodeh, Arnold Rosenberg, Jim Shearer, Larry Snyder, Clark Thompson and Les Valiant also deserve my thanks for their helpful remarks and criticisms. I am particularly thankful to Sandeep Bhatt and Margaret Lepley for proofreading preliminary drafts of the text.

In addition to the people who made direct contributions to the book, several individuals have, in the past, gone out of their way to further my educational development. Forman Acton, Len Adleman, Elva Aucklund, Alan Goldman, Charles Johnson, Phoebe Knipling, Dorothy Nelson, Frank Peterson and Dorothy Shriver stand out in this respect. I am especially indebted to Carl Hammer, David

Labovitz, Steve Maurer and Morris Newman for the many hours they spent teaching, advising and encouraging me.

Lastly, I would like to thank my wife Joanne for her support during the writing of the book and my entire family for the lifetime of sacrifices that they have made in order to insure that I obtained the best education available. Without their endless love and support, this book would not have been possible.

COMPLEXITY ISSUES IN VLSI

INTRODUCTION

During the past few years, there have been tremendous advances in very large scale integration (VLSI) fabrication technology. As a result, it is now possible to fabricate chips containing tens of thousands of transistors. In the near future, the number of transistors integrated onto a single chip will become even larger. In fact, chip capacities are expected to double every year or two for the next several years, and it is quite possible that fabrication of chips containing *millions* of transistors will be commonplace by the late 1980's [61].

The opportunites and challenges afforded by this rapidly-developing technology are formidable. On the one hand, the cost of computing has decreased dramatically. As a result, computation is now so cheap that chip-size micro-computers are routinely built into a multitude of products ranging from wrist watches to automobiles. On the other hand, the task of designing a state-of-the-art chip has become incredibly complex. And, as chip densities increase, so will the difficulty of designing a chip.

Motivated by the need to develop highly efficient, automated procedures for designing very large scale integrated circuits, theoretical researchers have been studying the underlying mathematical principles of large scale computation and circuit design. Thus far, their efforts have been directed largely towards answering the following three questions:

1) "What is a good model for VLSI chip design and computation?,"

2) "What communications networks can best perform important operations such as sorting, matrix multiplication and discrete Fourier transform?," and

3) "What is the best method for laying out a network on a chip?."

Initial progress on the first two questions is quite encouraging. A variety of good mathematical models have been proposed for chip design [20, 59, 92, 93], and several networks have been shown to be excellent structures for parallel computation [19, 21, 42, 43, 50, 64, 65, 66, 74, 75, 85, 87, 90, 94, 95]. The third question has proved to be more difficult, however, and is the subject of this book.

The book is divided into two parts with four chapters each. In Part I, we consider the layout problem for the *shuffle-exchange graph*. The shuffle-exchange graph is one of the best structures known for parallel computation. Among its many applications, a shuffle-exchange graph can be used to compute discrete Fourier transforms, multiply matrices, evaluate polynomials, perform permuations and sort lists [66, 87, 90]. The algorithms needed for these operations are quite simple and many require no more than logarithmic time and constant space per processor.

Because of its usefulness and simplicity, the shuffle-exchange graph serves as an excellent basis upon which to design and build chip-size microcomputers. One of the main difficulties with a shuffle-exchange architecture, however, is the problem of routing the wires which link the processors together in a shuffle-exchange network. Practical considerations dictate that the layout consume as little area as possible, have as few wire crossings as possible, and have as short wires as possible. This is because:

1) chips with large area cost more and experience lower yields than chips with small area,

2) chips with a large number of wire crossings (and, in particular, those with wires that cross many other wires) might have more problems with capacitive coupling than chips with a small number of crossings, and

3) chips with long wires are slower and harder to power than chips with short wires.

In [93], Thompson showed that any layout for the shuffle-exchange graph on a chip requires at least $\Omega(N^2/log^2N)$ area. Although a number of layouts for the shuffle-exchange graph have since been discovered [34, 89, 93], none achieves this asymptotic lower bound and none are practical. In Part I, we describe several new layouts for the shuffle-exchange graph. Those in Chapter 3 achieve the asymptotic lower bound for area while those in Chapter 4 appear to be suitable for practical applications. In addition, we prove that the layouts achieve the asymptotic lower bound for wire crossings and nearly achieve the asymptotic lower bound for maximum edge length.

In Part II, we consider the layout problem for a variety of other networks. We primarily concentrate on the problem of laying out the *mesh of trees* and the *tree of meshes*. Like the shuffle-exchange graph, the mesh of trees is an excellent network for parallel computation. In Chapter 6, we define the mesh of trees and show how it can be used to perform a variety of tasks (including sorting and matrix-vector multiplication) in a logarithmic number of steps. We also define the tree of meshes and show why its planar structure is fundemental to VLSI layout theory.

Unlike the shuffle-exchange graph, the mesh of trees and the tree of meshes are easy to lay out on a chip. The hard part rests in proving that their naive layouts are, in fact, optimal. To accomplish this task, we develop techniques for proving lower bounds on layout area, crossing number and edge length. The techniques are described in Chapters 7 and 8, where we prove that the N-node mesh of trees has layout area $O(Nlog^2N)$, crossing number $\Theta(NlogN)$ and maximum edge length $O(N^{1/2}logN/loglogN)$. In addition, we show that the N-node tree of meshes has layout area $O(NlogN)$. As a corollary, we find that the tree of meshes is the first counterexample to the conjecture that every planar graph has a linear-area layout.

The lower bound techniques developed in Chapters 7 and 8 can also be applied to a variety of other networks. For example, we prove in Chapter 7 that every network which computes an N-variable transitive function in time $T = o(N^{1/2})$ must have C wire crossings where $CT^2 \geq \Omega(N^2)$. This crossing-time tradeoff has many important applications and substantially generalizes Thompson's well-known area-time tradeoff [93]. Among other things, the tradeoff can be used to show that any layout of the N-node shuffle-exchange graph has at least $\Omega(N^2/log^2N)$ wire crossings.

We conclude the book with an addendum that summarizes the recent progress that has been made on layout-related problems. In some areas, the progress has been substantial, and forward pointers are supplied to several important papers.

PART I

LAYOUTS FOR THE SHUFFLE-EXCHANGE GRAPH

CHAPTER 1

FOUNDATIONS

In this chapter, we define the shuffle-exchange graph and the Thompson grid model of a chip. We also review the previous work on layouts for the shuffle-exchange graph. In Section 1.3, we describe Thompson's straightforward $O(N^2/log^{1/2}N)$-area layout for the N-node shuffle-exchange graph, and in Section 1.4, we describe Hoey and Leiserson's complex plane diagram. The complex plane diagram is useful for finding good layouts of the shuffle-exchange graph. For example, Hoey and Leiserson used the diagram to find an $O(N^2/logN)$-area layout for the N-node shuffle-exchange graph in [34]. In Chapter 2, we will use the diagram to find a variety of layouts for the N-node shuffle-exchange graph including one that requires only $O(N^2/log^{3/2}N)$ area. The complex plane diagram will also be used in Chapter 4 as an aid in the construction of good practical layouts for small shuffle-exchange graphs.

1.1 The Shuffle-Exchange Graph

The *shuffle-exchange graph* comes in various sizes. In particular, there is an N-node shuffle-exchange graph for every N which is a power of two. Each node of the $(N=2^k)$-node shuffle-exchange graph is associated with a unique k-bit binary string $a_{k-1}\cdots a_0$. Two nodes w and w' are linked via a *shuffle edge* if w' is a left or right cyclic shift of w (i.e., if $w = a_{k-1}\cdots a_0$ and $w' = a_{k-2}\cdots a_0 a_{k-1}$ or

$w' = a_0 a_{k-1} \cdots a_1$, respectively). Two nodes w and w' are linked via an *exchange edge* if w and w' differ only in the last bit (i.e., if $w = a_{k-1} \cdots a_1 0$ and $w' = a_{k-1} \cdots a_1 1$ or vice-versa). As an example, we have drawn the 8-node shuffle-exchange graph in Figure 1-1. Note that the shuffle edges are drawn with solid lines while the exchange edges are drawn with dashed lines. We shall follow this convention throughout the book.

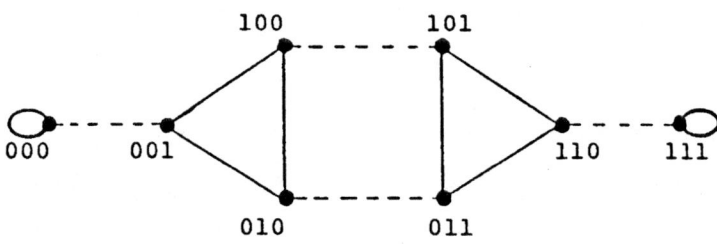

Figure 1-1: *The 8-node shuffle-exchange graph.*

By replacing the nodes and edges of the shuffle-exchange graph by processors and wires (respectively), the shuffle-exchange graph can be transformed into a very powerful parallel computer (which we call the *shuffle-exchange computer*). The computational power of the shuffle-exchange computer is partly derived from the fact that every pair of nodes in an N-node shuffle-exchange graph is linked by a path containing at most *2logN* edges and thus the communication time between any pair of processors is short.

More importantly, however, the shuffle-exchange computer is capable of performing a perfect shuffle on a set of data in a single parallel operation. For example, consider a deck of 8 cards distributed among the 8 processors of the 8-node shuffle-exchange graph so that processor *000* initially has card *0*, processor *001* initially has card *1*, processor *010* initially has card *2*, and so forth. Next, consider a (parallel) operation of the shuffle-exchange computer in which each processor $a_2a_1a_0$ sends its card across a shuffle edge to the neighboring processor $a_1a_0a_2$. It is easily verified that, after completion of the operation, processor *000* contains card *0* (the top card in the shuffled deck), processor *001* contains card *4* (the second card in the shuffled deck), and so forth.

The power of card shuffling and its mathematical abstractions are well known to magicians and mathematicians [23] as well as to computer scientists [87, 90]. For a good survey of the computational power of the shuffle-exchange graph, we recommend Schwartz' paper on ultracomputers [87]. In addition, Stone's paper [90] contains a nice description of some important parallel algorithms based on the shuffle-exchange graph.

1.2 The Thompson Grid Model

Among the many mathematical models that have been proposed for VLSI computation, the most widely accepted is due to Thompson and is known as the *Thompson grid model* [92, 93]. The grid model of a VLSI chip is quite simple. The chip is presumed to consist of a grid of vertical and horizontal *tracks* which are spaced apart by unit intervals. Processors are viewed as points and are located only at the intersection of grid tracks. Two layers of interconnect are used to route the wires. Vertical wires are routed in the top layer of interconnect and horizontal wires are routed in the bottom layer. Hence wires may cross but cannot overlap for any distance. Nor can wires overlap processors to which they are not adjacent. *Contact cuts*, which connect segments of the same wire which are in different layers, are also only located at the intersection of grid tracks. (The routing of wires in this fashion is also known as *layer per direction routing* and *Manhattan routing*.)

As an example, we have included a grid layout for the *8*-node shuffle-exchange graph in Figure 1-2. As before, the shuffle edges are drawn with solid lines while the exchange edges are drawn with dashed lines. Although we have not included them in the figure, contact cuts would be placed at points where a wire changes direction. Also notice that we have omitted the self-loops in Figure 1-2 since they are electrically redundant. In general, the processors need not all be placed on a single horizontal line (as they are in this example).

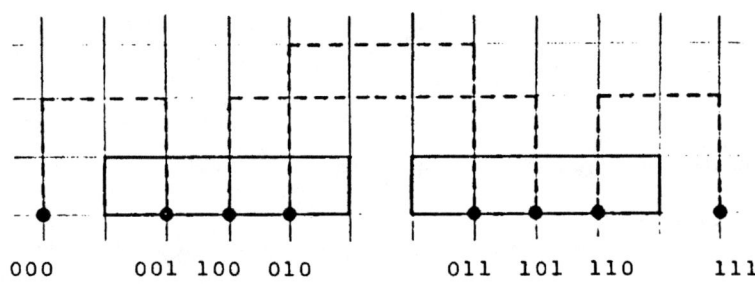

000   001 100 010   011 101 110   111

Figure 1-2: *A grid model layout of the 8-node shuffle-exchange graph.*

Practical considerations dictate that the area of a VLSI layout be as small as possible. The *area of a layout* in the grid model is defined to be the product of the number of horizontal tracks and the number of vertical tracks which contain a processor or wire segment of the layout. For example, the layout in Figure 2 has area *48*. As can be easily observed, this is far from optimal.

Other measures of interest are the wire area, crossing number, maximum edge length and maximum.edge crossing of a layout. The *wire area* of a layout is defined to be the sum of the lengths of the wires in the layout. The *crossing number* is the number of points in the layout where two wires cross. The *maximum edge length* is the length of the longest wire in the layout. The *maximum edge crossing* is the maximum number of points at which a single wire crosses other wires. For example, the layout in Figure 1-2 has wire area *50*, crossing number *8*, maximum edge length 9 and maximum edge crossing 4. The reader can verify that none of these values is optimal by skipping ahead to Figure 4-3.

1.3 An $O(N^2/log^{1/2}N)$-Area Layout

Thompson was the first to investigate VLSI layouts for the shuffle-exchange graph. In his thesis [93], he showed that *any* layout for the N-node shuffle-exchange graph requires at least $\Omega(N^2/log^2N)$ wire area. (We reprove this fact using crossing number arguments in Part II of the thesis.) In addition, he described a layout requiring only $O(N^2/log^{1/2}N)$ area. In what follows, we present Thompson's layout and give a simple proof that it does, in fact, require just $O(N^2/log^{1/2}N)$ area.

Given any k-bit string w, define the *size* of w to be the number of *1*-bits it contains. For example, the size of *10110* is *3*. Thompson's idea was to lay out the $N=2^k$ nodes of the shuffle-exchange graph on a straight line in order of nondecreasing size. It is easily seen that shuffle edges link nodes which have the same size and that exchange edges link nodes which have sizes differing by one. Thus the edges of such a layout are relatively short. In fact, nodes connected by shuffle edges can be placed in groups, so that only 2 horizontal tracks are used for all the shuffle connections. The remaining horizontal tracks are occupied by exchange edges.

The exchange edges are inserted from left to right so that each exchange edge occupies two vertical tracks and a portion of the lowest horizontal track which is empty at the time of its insertion. (For example, Figure 1-2 displays a layout for the 8-node shuffle-exchange designed in this way.) This well-known strategy for inserting exchange edges guarantees that the number of horizontal tracks used will be minimal, and equal to the maximum number of edges which must (at some fixed point) overlap one another. Since exchange edges link nodes which differ in size by one, it is easily seen that the maximum overlap is at most $O(\max\limits_{0 \le s \le k} B_s)$ where B_s is the number of nodes of size s.

It is easy to show that $B_s = C(k,s)$ for each s, where

$$C(k,s) = k!/[s!(k-s)!]$$

is the well-known function for binomial coefficients. It is also well-known that $C(k,s)$ achieves its maximum value at $s=k/2$ for any k. Using standard asymptotic analysis, it is easily shown that $C(k,k/2) \sim (2/\pi)^{1/2}(2^k/k^{1/2})$ for large k. (For a good review of such techniques, see Bender and Orszag's book [6].) Thus Thompson's layout requires only $O(N/log^{1/2}N)$ horizontal tracks. Since only 1 or 2 vertical tracks are needed to embed the vertical portions of the edges incident to any given node, we can conclude that Thompson's layout has area $O(N^2/log^{1/2}N)$.

1.4   The Complex Plane Diagram

In [34], Hoey and Leiserson observed that there is a very natural embedding of the shuffle-exchange graph in the complex plane. In what follows, we describe this embedding (henceforth referred to as the *complex plane diagram*) and point out some of its more important properties. In addition, we give a brief description of the method used by Hoey and Leiserson to transform the diagram into an $O(N^2/logN)$-area layout for the N-node shuffle-exchange graph.

1.4.1   Definition

Let $\delta_k = e^{2\pi i/k}$ denote the kth primitive root of unity. Given any k-bit binary string $w = a_{k-1} \cdots a_0$, let $p(w)$ be the map which sends w to the point

$$p(w) = a_{k-1}\delta_k^{k-1} + \cdots + a_1\delta_k + a_0$$

in the complex plane. As each node of the $(N=2^k)$-node shuffle-exchange graph corresponds to a k-bit binary string, it is possible to use the map to embed the shuffle-exchange graph in the complex plane. For example, we have done this for the *32*-node shuffle-exchange graph (whence $k=5$) in Figure 1-3. As usual, we have drawn the shuffle edges with solid lines and the exchange edges with dashed lines. For simplicity, each node is labeled with its value instead of its *5*-bit binary string. (By the *value* of a node, we mean the numerical value of the associated k-bit binary string. For example, the value of *01101* is *13*.)

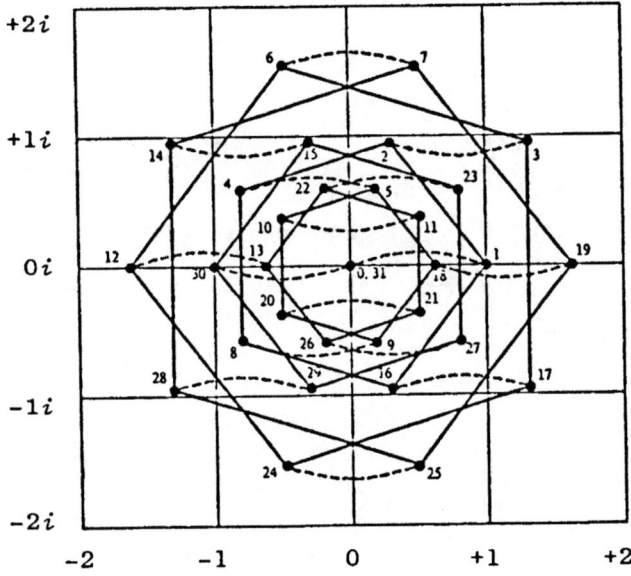

Figure 1-3: *The complex plane diagram for the 32-node shuffle-exchange graph. (Taken from* [34].)

1.4.2 Properties

Examination of Figure 1-3 indicates that the complex plane diagram has some very interesting properties. First, it is apparent that the shuffle edges occur in cycles (which we call *necklaces*) which are symmetrically placed about the origin.

This phenomenon is easily explained by the following identity:

$$\delta_k \, p(a_{k-1} \cdots a_0) = a_{k-1}\delta_k{}^k + a_{k-2}\delta_k{}^{k-1} + \cdots + a_1\delta_k{}^2 + a_0\delta_k$$

$$= a_{k-2}\delta_k{}^{k-1} + \cdots + a_0\delta_k + a_{k-1}$$

$$= p(a_{k-2} \cdots a_0 a_{k-1}).$$

Thus traversal of a shuffle edge corresponds to a $2\pi/k$ rotation in the complex plane.

Except for degenerate cases, the preceding identity also indicates that each necklace contains k nodes, each a cyclic shift of the other. Such necklaces are called *full necklaces*. *Degenerate necklaces* contain fewer than k nodes and, because they must have some symmetry, are mapped entirely to the origin of the complex plane diagram. For example, {*00000*} and {*0101, 1010*} are degenerate necklaces while both {*101, 011, 110*} and {*11100, 11001, 10011, 00111, 01110*} are full. As we note in the following lemma, the number of degenerate necklaces is quite small compared to the number of full necklaces.

Lemma 1·1: *There are* $O(N^{1/2})$ *degenerate necklaces and* $N/logN$ – $O(N^{1/2}/logN)$ *full necklaces in the N-node shuffle-exchange graph.*

Proof: A node w is in a degenerate necklace if its binary representation has a nontrivial symmetry with respect to cyclic shifts. Without loss of generality, such a string of bits must consist of a block of k/p bits which is repeated p times where p is some prime divisor of k. As there are $2^{k/p}$ binary strings of length k/p, this means that the number of nodes in degenerate necklaces is at most

$$\sum_{p \geq 2}^{p|k} 2^{k/p} \leq O(N^{1/2}).$$

The remaining N - $O(N^{1/2})$ nodes are in full necklaces. As each full necklace contains $logN$ nodes, there are $N/logN$ – $O(N^{1/2}/logN)$ full necklaces. ☐

It will often be convenient to refer to a necklace by one of its nodes. In particular, we will use the notation $\langle w \rangle$ to indicate the *necklace generated by w*. This is simply the collection of cyclic shifts of w. For example, the necklace generated by *101* is $\langle 101 \rangle = \{101, 011, 110\}$.

Exchange edges are also embedded in a very regular fashion in the complex plane diagram. In fact, each exchange edge is embedded as a horizontal line segment of unit length. This phenomenon is explained by the identity

$$p(a_{k-1} \ldots a_1 0) + 1 \;=\; a_{k-1}\delta_k{}^{k-1} + \ldots + a_1\delta_k + 1$$

$$= p(a_{k-1} \ldots a_1 1).$$

In some cases, several exchange edges are contained in the same horizontal line of the diagram. Such lines are called *levels*. For example, there are *9* levels in the diagram of the *32*-node shuffle-exchange graph shown in Figure 1-3. We will use the properties of levels in Chapter 2 to find an $O(N^2/log^{3/2}N)$-area layout for the N-node shuffle-exchange graph. They will also be used in Chapter 4 to find good practical layouts for small shuffle-exchange graphs.

1.4.3   An $O(N^2/logN)$-Area Layout

In [34], Hoey and Leiserson show how to use the complex plane diagram to construct an $O(N^2/logN)$-area layout for the N-node shuffle-exchange graph. Their method is quite complicated, however, and we have chosen not to include it here. The basic idea is to use the structural properties of the complex plane diagram to find an $O(N/log^{1/2}N)$-separator for the N-node shuffle-exchange graph whenever N is of the form 2^{2^r} for some $r \geq 0$. The separator can then used to construct an $O(N^2/logN)$-area layout by using Leiserson and Valiant's general layout technique for graphs with known separators. (Separators and their application to layouts are discussed in Part II.)

Shortly after writing [34], Hoey and Leiserson found a far simpler $O(N^2/logN)$-area layout for the N-node shuffle exchange graph which was, in addition, valid for all N. By the that time, however, we (as well as several others) had also observed that the complex plane diagram could be used to find a simple layout for the shuffle-exchange graph. This layout is described in Chapter 2.

CHAPTER 2

LAYOUTS BASED ON THE COMPLEX PLANE DIAGRAM

In this chapter, we present several layouts of the shuffle-exchange graph that are based on the complex plane diagram. We commence in Section 2.1 with a straightforward $O(N^2/logN)$-area layout of the N-node shuffle-exchange graph. As we mentioned in Chapter 1, this layout has also been discovered by many others (including Hoey and Leiserson). In Section 2.2, we show how the layout can be modified so as to require only $O(N^2/log^{3/2}N)$ area. The latter layout was also discovered independently by Steinberg and Rodeh [89]. We conclude the chapter by mentioning an additional $O(N^2/log^{3/2}N)$-area layout as well as a layout which might require even less area.

2.1 A Straightforward $O(N^2/logN)$-Area Layout

In this section, we describe a straightforward layout of the shuffle-exchange graph which requires only $O(N^2/logN)$ area. The layout is formed from a grid of levels and necklaces which we refer to as the *level-necklace grid*. Each row of the grid corresponds to a level of the complex plane diagram. The columns of the grid are divided into consecutive column pairs, each pair corresponding to a necklace. The leftmost column of each column pair corresponds to that part of the necklace which is contained in the left half of the complex plane. Similarly, the rightmost column of each pair corresponds to the part of the necklace contained in the right half of the complex plane.

The rows of the level-necklace grid must have the same top-to-bottom order as do the corresponding levels in the complex plane diagram. The columns, however, may be arranged arbitrarily (provided that columns corresponding to the same necklace are adjacent in the grid).

Each node of the shuffle-exchange graph is placed at the intersection of the row and column of the grid which correspond to the level and part of the necklace (left half or right half) to which it belongs in the complex plane diagram. For example, we have done this for a random ordering of the necklaces of the *32*-node shuffle-exchange graph in Figure 2-1. (Notice that we have used just one column each for the degenerate necklaces ⟨0⟩ and ⟨31⟩ since they each contain just one node. In general two columns will be required for necklaces which are mapped to the origin of the complex plane diagram, but the nodes of each such necklace should still be lumped togther at a single point of the level-necklace grid.)

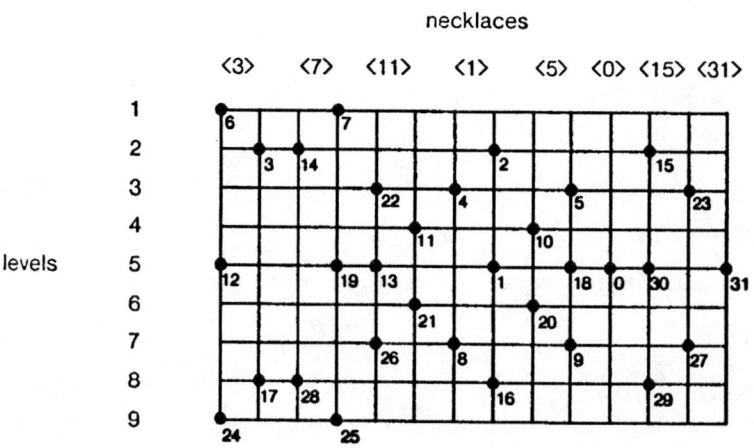

Figure 2·1: *A level-necklace grid for the 32-node shuffle-exchange graph.*

Given a level-necklace grid for a shuffle-exchange graph, it is not difficult to produce a layout for the graph. The main step is to partition the exchange edges in each row of the grid into nonoverlapping subsets. Each subset can then be assigned to a horizontal track of the layout. Except for the row corresponding to the real line in the complex plane diagram, the assignment of subsets to horizontal tracks within a row is arbitrary. (The assignment of horizontal tracks containing nodes on the real line must preserve the cyclic orientation of the nodes which are in necklaces that are mapped to the origin.)

Once this is done, the exchange edges can be inserted in the horizontal tracks and the shuffle edges can be inserted in the vertical tracks. (To be precise, some of the shuffle edges also occupy part of a horizontal track at the top or bottom of the layout.) By Lemma 1-1, the number of vertical tracks occupied by the necklaces is at most $2N/logN + O(N^{1/2})$. Since there are precisely $N/2$ exchange edges, at most $N/2 + 2$ horizontal tracks are contained in the layout. Thus the total area of the layout of the N-node shuffle-exchange graph is at most $N^2/logN + O(N^{3/2})$. As an example, we have displayed in Figure 2-2 a layout of the 32-node shuffle-exchange graph produced from the level-necklace grid in Figure 2-1.

2.2 An Improved $O(N^2/log^{3/2}N)$-Area Layout

It is possible to improve the layout described in Section 2.1 by reducing the number of horizontal tracks needed to embed the exchange edges. This can be done by reordering the necklaces from left to right so as to increase the average number of exchange edges which can be inserted on each horizontal track. For example, the ordering of the necklaces shown in Figure 2-3 results in far fewer horizontal tracks being used than did the ordering of necklaces shown in Figure 2-2.

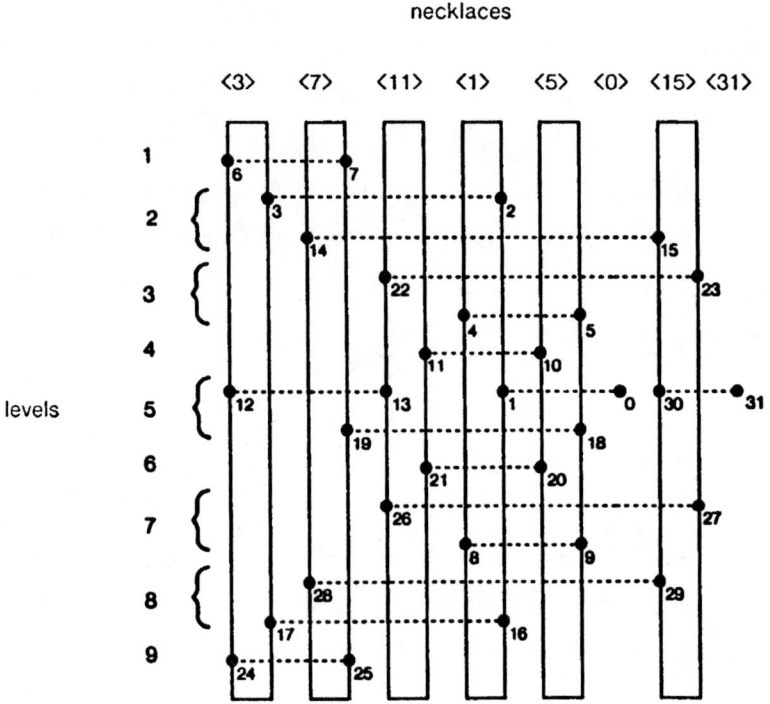

Figure 2-2: *Layout of the 32-node shuffle-exchange graph produced from the level-necklace grid shown in Figure 2-1.*

Although we do not know how best to order the necklaces in general, we have found several orderings which yield $O(N^2/log^{3/2}N)$-area layouts for the N-node shuffle-exchange graph. For instance, we will show in what follows that such a layout can be constructed by arranging the necklaces from left to right in order of nondecreasing size. (The *size of a necklace* is simply defined to be the size of any of its nodes.) As an example, the layout displayed in Figure 2-3 is of this form. (This observation has also been made by Steinberg and Rodeh in [89].)

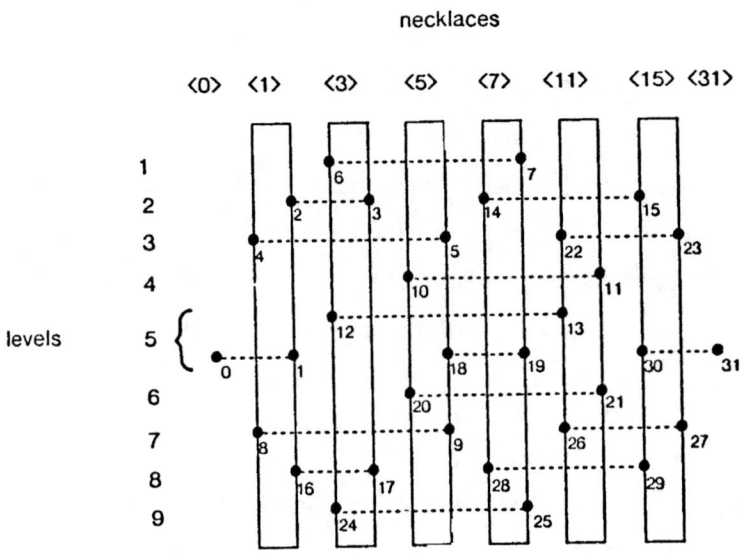

Figure 2-3: *An improved layout for the 32-node shuffle-exchange graph.*

In order to bound the number of horizontal tracks needed to insert the exchange edges, we will show that the maximum overlap of exchange edges *on each level* is at most the number of nodes of size $h = \lfloor(k-1)/2\rfloor$ on that level. Since the maximum overlap of exchange edges on each level is an upper bound on the number of horizontal tracks needed to insert the exchange edges on that level, we can thus conclude that the total number of horizontal tracks needed to insert all of the exchange edges is at most

$$B_h \leq B_{k/2} = (2/\pi)^{1/2} N / \log^{1/2} N + O(N/\log^{3/2} N).$$

Thus the resulting layout will have area at most

$$2(2/\pi)^{1/2} N^2 / \log^{3/2} N + O(N^2/\log^{5/2} N).$$

Although it is clear that the maximum *total* overlap (over all levels) of exchange edges is at most $B_{k/2}$, this is not sufficient to prove the result since any layout must also preserve the top-to-bottom partial order induced by the necklace structure on the exchange edges. It is only within individual levels that the top-to-bottom ordering of exchange edges is arbitrary. (As we noted earlier, some minor precautions are necessary for the level corresponding to the real line.) It is *not* immediately clear, however, why the maximum overlap on each level is at most the number of nodes of size $h = \lfloor (k-1)/2 \rfloor$ on that level. In what follows, we establish this result by breaking up each level into sublevels (for which the analysis is easier) and showing that the maximum overlap on each sublevel is at most the number of nodes of size h on that sublevel. The analysis requires some additional notation.

Consider a node of the form $a_{k-1} \cdots a_1 0$ for which either $a_{k-i}=0$ or $a_i=0$ or both for each $i \leq k$. We will refer to such a node as *basis node*. A node $b_{k-1} \cdots b_0$ is said to be *generated* by the basis node $a_{k-1} \cdots a_0$ if

1) $b_{k-i}=a_{k-i}$ and $b_i=a_i$ whenever $a_{k-i} \neq a_i$ for $1 \leq i \leq k-1$, and

2) $b_{k-i}=b_i$ whenever $a_{k-i}=a_i=0$ for $1 \leq i \leq k-1$.

For example, *10000* generates *10001, 11100* and *11101* but not *11111*.

It is not difficult to show that if u generates v, then both u and v are on the same level of the complex plane diagram. For example, let $u = a_{k-1} \cdots a_0$ and $v = b_{k-1} \cdots b_0$ and observe that

$$p(v) - p(u) = (b_{k-1} - a_{k-1}) \delta_k^{k-1} + \cdots + (b_1 - a_1) \delta_k + (b_0 - a_0)$$

$$= c_{k-1} \delta_k^{k-1} + \cdots + c_1 \delta_k + c_0$$

where $c_{k-i}=c_i$ for each i, $1 \leq i \leq k-1$. Since δ_k^{k-i} is the complex conjugate of δ_k^i for $1 \leq i \leq k-1$, we can conclude that $p(v) - p(u)$ is a real number and thus that u and v are in the same level of the complex plane diagram.

It is also easy to show that each node of the shuffle-exchange graph is generated by a unique basis node. In particular, the node which generates $b_{k-1} \cdots b_0$ can be found by

1) setting $b_0=0$ and (if k is even) setting $b_{k/2}=0$, and

2) setting $b_i=b_{k-i}=0$ for each i such that (originally) $b_i=b_{k-i}=1$.

Since exchange edges link nodes which have the same basis node, we can conclude from the preceding arguments that it is possible to partition each level of the complex plane diagram into *sublevels* so that the nodes in each sublevel are precisely the nodes generated by some basis node. We will now show that the maximum overlap on each sublevel is at most the number of nodes of size h on that sublevel.

Since the necklaces have been arranged from left to right in order of nondecreasing size, the overlap of exchange edges between two nodes of size s in any sublevel is at most $O(\max_{0 \leq s \leq k} B_s')$ where B_s' is the number of nodes in that sublevel with size s. In the following lemma, we compute B_s' and show that its maximum for any sublevel occurs at $s=h$.

Lemma 2-1: *Each basis node of size r generates B_s' nodes of size s, where*

1) $B_s' = C(h-r, i)$ *for $s=r+2i$ and $i \leq h-r$, and*

2) $B_s' = C(h-r, i)$ *for $s=r+2i+1$ and $i \leq h-r$*

when k is odd, and

1) $B_s' = C(h-r+1, i)$ *for $s=r+2i$ and $i \leq h-r+1$, and*

2) $B_s' = 2C(h-r, i)$ *for $s=r+2i+1$ and $i \leq h-r$*

when k is even.

Proof: When k is odd, there are precisely $h-r$ pairs $a_j=a_{k-j}=0$ in a basis node of size r. In order to generate a string of size $s=r+2i$ when k is odd, we must set $b_0=0$ and set i of the h-r pairs so that $b_j=b_{k-j}=1$. There are $C(h-r, i)$ such strings. To generate a string of size $s=r+2i+1$ when k is odd, we must set $b_0=1$ and choose i of the h-r pairs so that $b_j=b_{k-j}=1$. As before, there are

$C(h - r, i)$ such strings.

When k is even, there is also the degenerate pair $a_{k/2} = 0$. To generate a string of size $s = r + 2i$ when k is even, we must choose i of the $h - r + 1$ pairs so that $b_j = b_{k-j} = 1$ (this count includes the "pair" $b_0 = b_{k/2} = 1$). There are $C(h - r + 1, i)$ such strings. To generate a string of size $s = r + 2i + 1$ when k is even, we must set either $b_0 = 1$ and $b_{k/2} = 0$ or $b_0 = 0$ and $b_{k/2} = 1$, and choose i of the h-r pairs so that $b_j = b_{k-j} = 1$ $(j \neq k/2)$. There are $2C(h - r, i)$ such strings. $\square$

Given Lemma 2-1, it is easily checked that the maximum value of $B_s{}'$ for any sublevel (independent of the value of r) occurs when $s = h$. Thus the sum (over all sublevels) of the maximum overlap at each sublevel is at most the number of nodes of size $h = \lfloor (k-1)/2 \rfloor$ in the entire graph. This is at most $C(k, k/2) \sim (2/\pi)^{1/2}(2^k/k^{1/2})$. Thus the total area of the layout is no more than

$$2(2/\pi)^{1/2}N^2/\log^{3/2}N + O(N^2/\log^{5/2}N),$$

as claimed.

2.3 Other Layouts

It is not difficult to find other orderings of the necklaces which produce $O(N^2/\log^{3/2}N)$-area layouts for the N-node shuffle-exchange graph. For example, Lepley [51] used standard statistical methods to show that the arrangement of necklaces from left to right in order of nondecreasing radius produces such a layout. (By the *radius of a necklace*, we mean the radius of the circle in the complex plane which contains the necklace.) The proof is similar to the one in Section 2.2. In particular, it is shown that the maximum overlap in most levels occurs in the same place and that the total overlap of all of the levels at that point is $\Theta(N/\log^{1/2}N)$.

Although we consider it likely that better orderings of the necklaces exist, we do not know of any ordering which (provably) results in a layout with less than $o(N^2/log^{3/2}N)$ area. There is another ordering of interest, however. That is the ordering of the necklaces according to the minimum number represented by each necklace. (The *minimum number represented* by a necklace is simply the smallest value of any node in the necklace.) Coincidentally, the layout displayed in Figure 2-3 has such an ordering. Using techniques which are developed in Chapter 3, it is possible to show that the combined maximum overlap of exchange edges in all levels is at most $O(NloglogN/logN)$ for this ordering. This is substantially better than the $O(N/log^{1/2}N)$ overlap found in previous orderings and also very close to the lower bound of $\Omega(N/logN)$. Unfortunately, we do not know how to show that the maximum overlap at each level occurs in the same place. In fact, it appears that this may not be the case. (We are deeply indebted to Kleitman for pointing out the possibility of such an improvement. Although we were not able use his idea in the context of complex plane diagram layouts, it was crucial to the development of the asymptotically optimal layout described in Chapter 3.)

For orderings which have a small combined maximum overlap but for which the maximum overlap at each level is difficult to compute (such as the ordering by minimum value represented), it may be possible to improve the situation by altering the level structure. As Miller pointed out to us, there are many possible levelings of the exchange edges. (By a *leveling*, we mean any arrangement of the exchange edges in levels which is consistent with the necklace structure of the complex plane diagram.) Although we have investigated several levelings, we have not found any (provably) better layouts for the shuffle-exchange graph by this method.

CHAPTER 3

MORE SOPHISTICATED LAYOUTS

In Section 3.3 of this chapter, we describe an asymptotically optimal $O(N^2/log^2N)$-area layout for the N-node shuffle-exchange graph. Unlike the previously described layouts, the optimal layout is fairly sophisticated and requires a substantial amount of preliminary machinery. Most of the necessary definitions and lemmas are included in Section 3.1. In Section 3.2, we describe and analyze a near-optimal preliminary version of the optimal layout. The optimal layout is then described in Section 3.3. In Section 3.4, we extend the methods developed in earlier sections in order to show that certain useful supergraphs of the N-node shuffle-exchange graph can also be laid out in $O(N^2/log^2N)$ area. We have also included an appendix to the chapter in which we prove Lemmas 3-1 through 3-4.

3.1 Preliminaries

The layouts described in this chapter are based on some important combinatorial properties of strings which contain long blocks of consecutive zeros. Before describing the layouts, however, it is useful to review some of these properties. In this section, we mention several combinatorial lemmas and definitions that will be heavily used in the analysis which follows later. As the proofs of the lemmas are somewhat complicated, they have been included in the appendix.

In what follows, we will be particularly interested in the size and location of the longest block of consecutive 0-bits in the k-bit binary string associated with each node. In order that the size of this block be the same for all nodes within a necklace, we allow blocks to begin at the end and end at the beginning of a string. For example, the longest block of zeros in the string *01010* starts at the fifth bit and has length two.

Let $\Psi_k(t)$ denote the number of k-bit strings for which the longest block of consecutive zeros has length t. For example, $\Psi_3(2) = 3$. The following combinatorial lemma provides a good asymptotic bound on the growth of $\Psi_k(t)$.

Lemma 3-1: *For* $(\log k)/2 + \log \ln k \leq t \ll k$ *and* $k \to \infty$,

$$\Psi_k(t) \sim 2^k \left(e^{-k2^{-(t+2)}} - e^{-k2^{-(t+1)}} \right).$$

In order to illustrate the important features of the function in Lemma 3-1, we have sketched a graph of $2^{-k}\Psi_k(t)$ versus t in Figure 3-1.

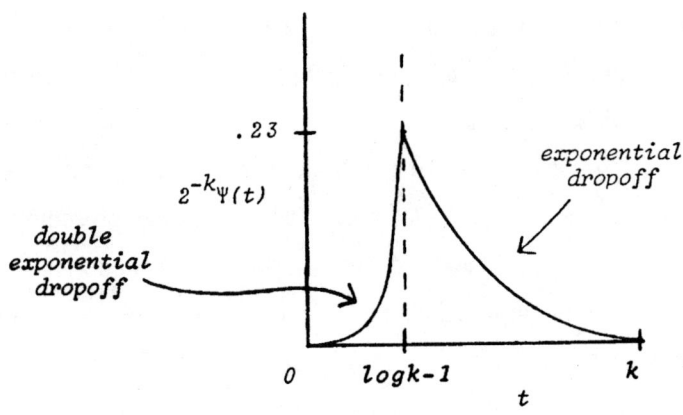

Figure 3-1: *Density of k-bit binary strings for which the longest block of consecutive zeros has length t.*

The maximum of $2^{-k}\Psi_k(t)$ occurs at $t = logk-1$ whence

$$2^{-k}\Psi_k(t) = (e^{1/2} - 1)/e$$

$$\approx .23865 .$$

For $t > logk - 1$, $2^{-k}\Psi_k(t)$ decreases *exponentially* as t increases. For $t \leq logk - 1$, $2^{-k}\Psi_k(t)$ decreases *doubly exponentially* as t decreases.

Roughly speaking, Lemma 3-1 states that the longest block of consecutive zeros in nearly *1/4* of all *k*-bit strings has length precisely *logk - 1*. Further, there are not many strings of length *k* with substantially more than *logk* consecutive zeros and even fewer strings for which the longest block of consecutive zeros has length substantially less than *logk*. This information is further quantified in the following lemma.

Lemma 3-2:   *The number of k-bit strings for which the longest block of consecutive zeros has length less than logk - loglnk - 1 or length greater than 2logk is at most* $O(2^k/k) = O(N/logN)$.

Leiserson observed in [56] that any M wires can be added to a layout by inserting at most $2M$ vertical and $2M$ horizontal tracks. Hence M wires can be added to a $\Omega(M)$-by-$\Omega(M)$ layout without increasing the area by more than a constant factor. As any layout for the N-node shuffle-exchange graph must have $\Omega(N/logN)$ vertical and $\Omega(N/logN)$ horizontal tracks, the preceding observation means that a nearly complete layout for the shuffle-exchange graph with area A can be extended to a complete layout with area $O(A)$. This result will be used at several points in the book and is stated formally in the following proposition.

Proposition 3-1:   *Any area A layout that contains all but $O(N/logN)$ nodes and edges of the N-node shuffle-exchange graph can be extended to form a complete layout for the N-node shuffle-exchange graph with area $O(A)$.*

As an immediate application of Proposition 3-1 and Lemma 3-2, we can henceforth ignore nodes for which the longest block of zeros has length less than $logk - loglnk - 1$ or length greater than $2logk$. Such nodes can always be added later with the addition of $O(N/logN)$ vertical and horizontal tracks. Similarly, Proposition 3-1 and Lemma 1-1 imply that we can ignore nodes that are contained in degenerate necklaces.

We will also be interested in the size of the second longest block of consecutive zeros in each string. Usually, the size of the second longest block of zeros will be very close to the size of the longest block of zeros. We state this observation more precisely in the following lemma.

Lemma 3-3: *The sum over all necklaces of the difference in length between the longest and second longest blocks of consecutive zeros is at most* $O(N/logN)$.

Using information about the size and location of blocks of zeros within a necklace, it is possible to distinguish one particular node in the necklace. More precisely, we define the *distinguished node of a necklace* to be the node containing the longest leading block of zeros. For example, 00101 is the distinguished node of $\langle 01010 \rangle$. Should two or more nodes of a necklace begin with equal and maximum length blocks of zeros, then each node of the necklace contains at least two blocks of zeros of maximum length. In such cases, we distinguish that node for which the leading block of zeros is maximum and for which the second occurence of a maximum length block of zeros is as near as possible to the beginning of the string. For example, 01011 (not 01101) is the distinguished node of the necklace $\langle 10101 \rangle$. For some necklaces, such as $\langle 111 \rangle$ and $\langle 1010101 \rangle$, there is no uniquely distinguished node. As we show in the following lemma, such necklaces are sufficiently rare that (by Proposition 3-1) we need not consider them further.

Lemma 3-4: *At most* $O(N/logN)$ *nodes are contained in necklaces which fail to have a uniquely distinguished node.*

We refer to the leading block of zeros of a distinguished node as the *primary block of zeros*. If a distinguished node has two or more maximum length blocks of zeros, then the maximum length block following the primary block is referred to as the *secondary block of zeros*. These definitions can be easily extended to any node contained in a necklace which has a uniquely distinguished node. For example, the primary block of zeros of *01010* starts in the fifth bit and has length two. Note that this string does *not* have a secondary block of zeros. As another example, we note that the secondary block of zeros in the string *11010* consists solely of the fifth bit. Note that the secondary block of zeros (if it exists) always has the same length as the primary block of zeros.

If the last bit of a node occurs in the primary block of zeros, we call that node a *primary node*. Similarly, if the last bit of a node occurs in the secondary block of zeros, we call the node a *secondary node*. For example, *10110* is a primary node, *11010* is a secondary node and *10010* is neither primary nor secondary.

Note that all primary and secondary nodes are necessarily even. (We say that a node is *even* if its last bit is *0* and *odd* if its last bit is *1*.) Note also that, by Lemma 3-2 and Proposition 3-1, we need only consider necklaces which contain between $logk - loglnk - 1$ and $2logk$ primary nodes. Such necklaces will also have at most $2logk$ secondary nodes.

In what follows, we will represent nodes in terms of their corresponding distinguished nodes. More precisely, we use the notation $a_{k-1}\cdots a_{i+1}\overline{a_i}a_{i-1}\cdots a_0$ to denote the node $a_{i-1}\cdots a_0 a_{k-1}\cdots a_i$. For example, *00101* denotes the node *10010*. Using this notation, a primary node has the form $0\cdots\overline{0}\cdots 0w$ while a secondary node has the form $0\cdots 0w'0\cdots\overline{0}\cdots 0w''$ where $0\cdots 0w$ and $0\cdots 0w'0\cdots 0w''$ are assumed to be distinguished nodes.

3.2 A Near-Optimal Layout

We are now prepared to describe a near-optimal preliminary version of the optimal layout. In Section 3.3, we will show how to modify this layout in order to construct an optimal $O(N^2/log^2 N)$-area layout for the N-node shuffle-exchange graph.

3.2.1 Location of the Nodes

The near-optimal layout is constructed from a $logN$ x $O(N/logN)$ grid of nodes. Each column of the grid corresponds to a necklace of the shuffle-exchange graph. The nodes of each necklace are ordered from top to bottom so that the ith node is a left cyclic shift of the $(i-1)st$ node for each i and so that the distinguished node is placed in the bottom row. The necklaces are ordered from left to right so that the values of the distinguished nodes form an increasing sequence. For example, we have constructed such a grid for the 32-node shuffle-exchange graph in Figure 3-2. In the figure, we have represented each node in terms of the associated distinguished node. This representation readily illustrates the fact that the last bit of any node in the ith row corresponds to the ith bit of the associated

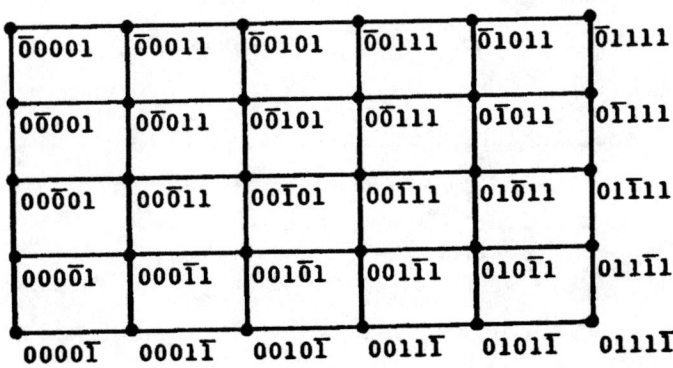

Figure 3-2: *The grid of nodes for the 32-node shuffle-exchange graph.*

distinguished node. Note that the necklaces $\langle 00000 \rangle$ and $\langle 11111 \rangle$ have not been included since they are degenerate.

3.2.2   Insertion of the Edges

It is easily observed that the shuffle edges can be inserted in the grid with the addition of $O(N/logN)$ vertical and 2 horizontal tracks. In the following, we will show that the exchange edges can be inserted with the addition of $O(NloglogN/logN)$ vertical and horizontal tracks. Thus the total area of the layout will be $O(N^2(loglogN)^2/log^2N)$. This is only a factor of $O((loglogN)^2)$ off from the lower bound of $O(N^2/log^2N)$.

The analysis is divided into two parts. In part (a), we show that only $O(NloglogN/logN)$ exchange edges link nodes which are in *different* rows of the grid. Thus such edges can be inserted with the addition of at most $O(NloglogN/logN)$ vertical and horizontal tracks. In part (b), we conclude the analysis by showing that at most $O(N/logN)$ horizontal tracks are needed to insert the exchange edges that link two nodes in the *same* row.

(a)   Exchange Edges That Link Nodes in Different Rows

Consider an exchange edge that links two nodes which are in different rows of the grid. In particular, assume that the edge is incident to an even node in the *ith* row for some *i*. By definition, the even node can be represented as $w\bar{0}w'$ where $|w| = i\text{-}1$ and $w0w'$ is the distinguished node of $\langle w0w' \rangle$. The exchange edge is also incident to the odd node $w\bar{1}w'$. By assumption, $w\bar{1}w'$ is not located in the *ith* row and thus $w1w'$ is *not* a distinguished node. Since $w0w'$ *is* a distinguished node, we know that the *ith* bit of $w0w'$ (the bit that was changed in order to produce $w\bar{1}w'$) must be in the primary or secondary block of zeros of $w0w'$. Otherwise, the primary and (if it exists) secondary blocks of zeros of $w1w'$ would be identical in location and size to the primary and secondary blocks of zeros of $w0w'$.

This would imply that w/w' is also distinguished, a contradiction. Thus $w\overline{0}w'$ must be a primary or secondary node. As was previously mentioned, we can assume that each necklace has at most $2logk = 2loglogN$ primary and $2loglogN$ secondary nodes. Thus at most $4loglogN$ nodes in each necklace are both even and incident to an exchange edge that links nodes in different rows. Since every exchange edge is incident to an even node and since there are $O(N/logN)$ necklaces, we can conclude that there are at most $O(NloglogN/logN)$ exchange edges that link nodes in different rows.

(b)   Exchange Edges That Link Nodes in the Same Row

We next show that those exchange edges which link two nodes that are in the same row can be inserted with the addition of at most $O(N/logN)$ horizontal tracks. Once again, the analysis is divided into two parts. In the first part, we show that at most $O(N/logN)$ exchange edges are contained in the first $logk$ rows. Such edges can be trivially inserted with the addition of $O(N/logN)$ horizontal tracks. In the second part, we show that only 2^{k-i} horizontal tracks are needed to insert the exchange edges in the ith row for any $i > logk$. Since

$$\sum_{i=logk+1}^{k} 2^{k-i} \leq 2^k/k = N/logN,$$

this will be sufficient to show that at most $O(N/logN)$ additional horizontal tracks are necessary to insert the remaining exchange edges.

Consider a necklace which has t primary nodes for some $t \leq logk$. By definition, the nodes in the first t rows of such a necklace are all even. Thus, such a necklace can have at most $r = logk - t$ odd nodes in the first $logk$ rows. By Lemma 3-1, we know that there are

$$\Psi_k(t)/k \sim (2^k/k)(e^{-k2^{-t-2}} - e^{-k2^{-t-1}})$$

such necklaces for $(logk)/2 + loglnk \leq t \ll k$. By Lemma 3-2, we can assume that $t \geq logk - loglnk - 1$ and thus the total number of odd nodes occurring in the first *logk* rows is at most

$$\sim \sum_{t=logk-loglnk-1}^{logk} (logk - t)(2^k/k)(e^{-k2^{-t-2}} - e^{-k2^{-t-1}})$$

$$= (2^k/k) \sum_{r=0}^{loglnk+1} r(e^{-k2^{r-2-logk}} - e^{-k2^{r-1-logk}})$$

$$= (2^k/k) \sum_{r=0}^{loglnk+1} r(e^{-2^{r-2}} - e^{-2^{r-1}})$$

$$\leq (2^k/k) \sum_{r=1}^{loglnk+1} e^{-2^{r-2}}$$

$$\leq (2^k/k) \sum_{r=1}^{\infty} e^{-2^{r-2}}$$

$$= O(N/logN).$$

Since every exchange edge is incident to an odd node, the above bound implies that at most $O(N/logN)$ exchange edges are contained in the first *logk* rows.

We next consider the number of horizontal tracks necessary to insert the exhange edges contained in the *ith* row for $i > logk$. This number is identical to the maximum number of exchange edges that can overlap each other at a single point of the *ith* row. In Figure 3-3, we illustrate the necessary conditions for two exchange edges to overlap in the *ith* row. All representations are in terms of distinguished nodes.

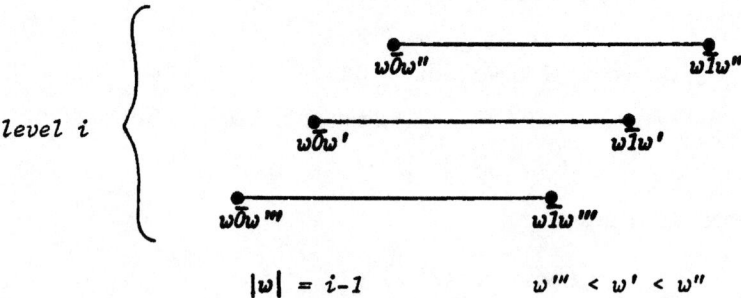

Figure 3-3: *Necessary conditions for exchange edges to overlap in the ith row.*

Note that the even end of an exchange edge is always to the left of the odd end. Also note that any node which occurs between $w\bar{0}w'$ and $w\bar{1}w'$ must be represented as $w\bar{0}w''$ where $w''{>}w'$ or as $w\bar{1}w'''$ where $w'''{<}w'$. In either case, the exchange edge incident to the overlapped node extends beyond the exchange edge linking $w\bar{0}w'$ to $w\bar{1}w'$. Since there are at most $2^{k-i} - 1$ nodes between $w\bar{0}w'$ and $w\bar{1}w'$, these facts imply that at most 2^{k-i} exchange edges can overlap at any point of the ith row. This observation completes the argument that the near optimal layout requires only $O(N^2(loglogN)^2/log^2N)$ area.

3.3   An Optimal $O(N^2/log^2N)$-Area Layout

In this section, we modify the layout described in Section 3.2 in order to produce an optimal $O(N^2/log^2N)$-area layout for the N-node shuffle-exchange graph. In particular, we will show how to relocate the primary and secondary nodes of each necklace so that they are closer to and in the same row as the nodes to which they are linked via an exchange edge. Before going into the details of this relocation, however, it is necessary to introduce some additional terminology.

3.3.1   More Definitions

In order to construct an optimal layout for the shuffle-exchange graph, we have found it necessary to break up each necklace into two or, possibly, three pieces. The *basic piece* of each necklace consists of all those nodes which are neither primary nor secondary. The *primary piece* of each necklace consists of the primary nodes while the *secondary piece* consists of the secondary nodes (if there are any). For example, the basic piece of $<01011>$ is $\{0\bar{1}011, 010\bar{1}1, 0101\bar{1}\}$, the primary piece is $\{\bar{0}1011\}$, and the secondary piece is $\{01\bar{0}11\}$.

It is also necessary to extend the notion of a distinguished node to include pieces of necklaces. The *distinguished node of a basic piece* is the same as the distinguished node of the associated necklace. The *distinguished node of a primary piece* of a necklace is that node of the necklace which becomes distinguished when we ignore the primary block of zeros (i.e., when we temporarily replace the primary block of zeros in each node of the necklace with an equal-length block of ones). Similarly, the *distinguished node of a secondary piece* of a necklace is that node which becomes distinguished when we ignore the secondary block of zeros. For example, *010110111* is the distinguished node of the basic piece of ⟨*010110111*⟩, *011011101* is the distinguished node of the primary piece, and *011101011* is the distinguished node of the secondary piece. Note that the distinguished nodes of the primary and secondary pieces of any necklace are necessarily odd nodes and thus are contained in the basic piece of the necklace.

It is important to note that some necklaces (such as ⟨*01111*⟩) have a distinguished node but do not have a distinguished node for the primary or secondary piece of the necklace. Fortunately, arguments such as those used to prove Lemmas 3-3 and 3-4 can be used to show that at most O($N/logN$) nodes are contained in such necklaces. Thus (by Proposition 3-1), we can assume henceforth that every piece of every necklace has an associated distinguished node.

3.3.2   Location of the Nodes

As in Section 3.2, the layout is constructed from a $logN$ x O($N/logN$) grid of nodes. Each column of the grid corresponds to a piece of a necklace. The nodes of each piece are arranged within a column so that a node of the form $a_{k-1} \cdots \overline{a}_{k-i} \cdots a_0$ (where $a_{k-1} \cdots a_0$ is assumed to be the distinguished node of the associated piece) is placed in the ith row of the grid. Note that nodes in the basic piece of any necklace (these include all odd nodes) are in the same row as they were in the near-optimal layout described in Section 3.2. The columns are ordered from left to right so that the values of the distinguished nodes of the associated pieces form a nondecreasing sequence. For example, we have constructed such a grid for $k=5$ in Figure 3-4.

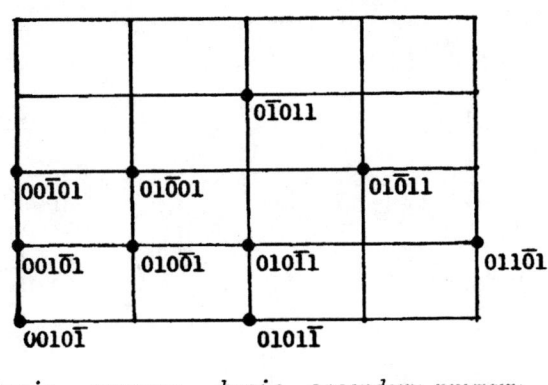

Figure 3-4: *Relocated nodes for the 32-node shuffle-exchange graph.*

Note that the necklaces ⟨*00001*⟩, ⟨*00011*⟩, ⟨*00111*⟩, and ⟨*01111*⟩ have not been included in Figure 3-4 since their associated primary pieces do not have distinguished nodes.

3.3.3 Insertion of the Edges

As each necklace is broken up into at most four *contiguous* pieces in the modified grid (the basic piece may have been broken up into two contiguous pieces), the shuffle edges can be inserted with the addition of at most $O(N/logN)$ vertical and horizontal tracks. In what follows, we will show that at most $O(N/logN)$ vertical and horizontal tracks are needed to insert all of the exchange edges as well. Thus the area of the layout will be $O(N^2/log^2N)$, which is optimal.

As before, we divide the analysis of the exchange edges into two parts. We first show that at most $O(N/logN)$ exchange edges link nodes which are in different rows of the grid. Such edges can thus be trivially inserted with the addition of at most $O(N/logN)$ vertical and horizontal tracks. We then show that those exchange edges which link two nodes in the same row can be inserted with the addition of only $O(N/logN)$ horizontal tracks. The arguments will be very similar to those in Section 3.2.2.

(a)   Exchange Edges That Link Nodes in Different Rows

Consider an exchange edge that links two nodes which are in different rows of the grid. Since only primary and secondary nodes have been relocated, we can conclude from the arguments of Section 3.2.2a that the even node which is incident to the edge is either a primary or secondary node. In what follows, we will show that the even node is, in fact, a primary node.

Assume for the purposes of contradiction that the even node is a secondary node. Then this node can be represented as $w\bar{0}w'$ where $w0w'$ is the distinguished node of the secondary piece of $\langle w0w'\rangle$ and $|w|=i{-}1$ for some i. By definition, $w\bar{0}w'$ is located in the ith row of the grid and is linked to $w\bar{1}w'$ via the exchange edge. Since $w\bar{1}w'$ is odd, it is contained in the basic piece of $\langle w1w'\rangle$. By assumption, $w\bar{1}w'$ is not also in the ith row and thus $w1w'$ cannot be the distinguished node of $\langle w1w'\rangle$. Since the lengths of the two blocks of zeros in $w1w'$ created by switching the ith bit from 0 to 1 are less than the length of the primary block of zeros (in fact, the sum of their lengths is precisely one less than the length of the primary block), $w1w'$ will be the distinguished node of $\langle w1w'\rangle$ precisely when $w0w'$ is the node distinguished in $\langle w0w'\rangle$ by ignoring the secondary block of zeros. By definition, this is the case precisely when $w0w'$ is the distinguished node of the secondary piece of $\langle w0w'\rangle$. By assumption, $w0w'$ *is* the distinguished node of the secondary piece of $\langle w0w'\rangle$ and thus we can conclude that $w1w'$ is the distinguished node of $\langle w1w'\rangle$, a contradiction.

Next consider a *primary* node which is incident to an exchange edge linking two nodes in different rows of the grid. By the preceding arguments, this node must be of the form $w l \overbrace{0 \cdots 0}^{l_1} 0 \overbrace{0 \cdots 0}^{l_2} 1 w'$ where $w l 0 \cdots 0 1 w'$ is the distinguished node of the primary piece of $\langle w l 0 \cdots 0 1 w' \rangle$ and either l_1 or l_2 is larger than or equal to the length of the longest block of zeros in $w l l w'$. Otherwise, $w l \overbrace{0 \cdots 0}^{l_1} 1 0 \overbrace{0 \cdots 0}^{l_2} 1 w'$ would (by definition) be the distinguished node of $\langle w l \overbrace{0 \cdots 0}^{l_1} 1 0 \overbrace{0 \cdots 0}^{l_2} 1 w' \rangle$ and thus $w l \overbrace{0 \cdots 0}^{l_1} 1 0 \overbrace{0 \cdots 0}^{l_2} 1 w'$ would be on the same row as $w l \overbrace{0 \cdots 0}^{l_1} 0 0 \overbrace{0 \cdots 0}^{l_2} 1 w'$ a contradiction. Each necklace contains at most $2r$ such primary nodes where r is the difference between the lengths of the longest and second longest block of zeros in any string of the necklace. By Lemma 3-3, we can conclude that there are at most $O(N/logN)$ such primary nodes in the entire shuffle-exchange graph. Thus, at most $O(N/logN)$ exchange edges link nodes which are in different rows.

(b)   Exchange Edges That Link Nodes in the Same Row

Using the analysis developed in Section 3.2.2b, it is not difficult to show that at most $O(N/logN)$ horizontal tracks are needed to insert the exchange edges which link two nodes that are in the same row. In particular, there are still only $O(N/logN)$ odd nodes in the top $logk$ rows of the grid and thus at most $O(N/logN)$ exchange edges are contained in the top $logk$ rows. These can be trivially inserted with the addition of just $O(N/logN)$ horizontal tracks.

Again following the methods of Section 3.2.2b, it is not difficult to show that two exchange edges overlap on the ith row only if the first i bits of the associated nodes are identical. Thus at most 2^{k-i} tracks are needed to insert all of the exchange edges in the ith row for all $i>logk$. Summing, we can again conclude that at most $O(N/logN)$ additional horizontal tracks are needed to insert the remaining exchange edges.

3.3.4   Comments

The methods developed in this chapter can be used to find several other
optimal layouts for the shuffle-exchange graph. The key variant is the method by
which a node is distinguished. In particular, this method must be impervious to
small alterations in the necklace. (This is so that most exchange edges will link
nodes which are in the same row of the grid.) Only by changing the value of a bit
in a small segment of the necklace (such as in the primary or secondary block of
zeros) should we be able to globally change the distinguished node.

Another method of distinguishing a node is to select that node in the necklace
which has the minimal value. Although the proof is very difficult, it can be shown
that the layout for the N-node shuffle-exchange graph constructed in this manner
has at most $O(N^2/log^2N)$ area. In the following section we will describe additional
methods of distinguishing nodes.

In Chapter 7, we will show that every layout for the shuffle-exchange graph
must have $\Omega(N^2/log^2N)$ wire crossings. The layouts just described have precisely
this many crossings (since they have at most this much area) and thus they are
optimal with respect to crossing number. They are *not known* to have optimal
maximum edge length, however. In Part II, we show that every layout of the N-
node shuffle-exchange graph must have some edge of length at least $\Omega(N/log^2N)$.
All the layouts we have considered thus far contain wires of length $O(N/logN)$.

3.4   Layouts With Additional Edges

For some applications (such as the calculation of the discrete Fourier
transform), it is useful to consider networks which have more than just shuffle and
exchange edges. In particular, we are interested in layouts for the shuffle-exchange
graph which also include shift, reverse and transpose edges. In what follows, we

show how to modify the optimal layout for the shuffle-exchange graph so that these additional edges can be inserted without increasing the total area by more than a constant factor.

3.4.1 Shift Edges

Shift edges link the *i*th node to the $(i+1)$st node for all *odd i.* When combined with the exchange edges, the resulting network will have links between the *i*th and the $(i+1)$st nodes for all *i.* The inclusion of such edges facilitates the computation of discrete Fourier transforms at sequential intervals of a continuous signal. In such applications, the input data contained in the *i*th processor is shifted to the $(i+1)$st processor for each *i* after each computation of a discrete Fourier transform. The graph consisting of shuffle, exchange and shift edges is known as the *shuffle-shift graph.*

Using the methods developed in Section 3.3, it is not difficult to show that the *N*-node shuffle-shift graph can be laid out using only $O(N^2/log^2 N)$ area. As before, the necklaces are broken into two or three pieces and placed in a grid according to the value of the associated distinguished node. Thus the shuffle edges can be inserted as before using only $O(N/logN)$ vertical and horizontal tracks.

For most odd nodes, adding a *1* to the value of the node changes only a relatively small number of bits at the end of the string. Thus it can be shown that at most $O(N/logN)$ shift edges link nodes which are in different rows. These can be easily inserted using only $O(N/logN)$ vertical and horizontal tracks. Of those edges which link nodes in the same row, at most $O(N/logN)$ are contained in the first *logk* rows. For *i>logk*, at most 2^{k-i} shift edges overlap at any point of the *i*th row. By introducing an extra vertical track for each necklace piece, it is possible to separate the layout of the shift edges on each level from that of the exchange edges. Thus both can be inserted simultaneously in the *i*th row using only $O(2^{k-i})$

total horizontal tracks. By the arguments of Section 3.3, this means that at most $O(N/logN)$ additional horizontal tracks are needed to embed all of the remaining shift and exchange edges, thus completing the argument.

3.4.2   Reverse Edges

Reverse edges link pairs of nodes that are associated with binary strings which are reverses of each other. For example, $a_{k-1} \cdots a_0$ is linked to $a_0 \cdots a_{k-1}$ via a reverse edge. Since the algorithm which computes discrete Fourier transforms on the shuffle-exchange network leaves the output for node $a_{k-1} \cdots a_0$ in node $a_0 \cdots a_{k-1}$, reverse edges provide a fast and convenient way of straightening out the solution. The graph consisting of shuffle, exchange, shift and reverse edges is called the *shuffle-shift-reverse graph.*

Using the techniques developed in Section 3.3, it is also possible to show that the N-node shuffle-shift-reverse graph can be laid out in $O(N^2/log^2N)$ area. The basic idea is to modify the layout described in Section 3.4.1 so that

1) pieces of necklaces which are reverses of each other are paired together in the left-to-right ordering, and

2) pieces of necklaces are folded in half.

The first constraint insures that the maximum overlaps of the reverse edges in each row are small while the second constraint insures that most reverse edges link nodes which are in the same row. Although it is not immediately obvious, it can be checked that these modifications do not substantially change the procedure for inserting the shuffle, shift and exchange edges which was described in Section 3.4.1. Thus all of the edges can be inserted using at most $O(N/logN)$ vertical and horizontal tracks.

3.4.3   Transpose Edges

Transpose edges link the *i*th node to the $(N-1-i)$th node for each *i*. Viewed in terms of binary strings, transpose edges link each node to its complement. Although we do not know of any specific applications of transpose edges, they would be useful for problems that require frequent transposition of the data.

By further modifying the optimal layout for the shuffle-shift-reverse graph, it is possible to add transpose edges without increasing the total area by more than a constant factor.   In particular, the layout should be modified so that

1) pieces of necklaces which are complements of each other are paired together in the left-to-right ordering, and

2) the distinguished node is selected on the basis of the location of the longest block of consecutive identical bits (be they zeros *or ones*).

The first constraint insures that the maximum overlaps of the transpose edges in each row are small while the second constraint insures that most transpose edges link nodes which are on the same row. Although we do not present the details here, it is possible to show that such a layout can be constructed using only $O(N^2/log^2N)$ area, the least possible.

Appendix: Proofs of Lemmas 3-1 Through 3-4

We now present the proofs of Lemmas 3-1 through 3-4. Such results can also be found in the recent work of Guibas and Odlyzko [31, 32]. We are deeply indebted to Kleitman for suggesting the proof of Theorem 3-1.

In what follows, we will write $\overline{\Psi}_k(t)$ to denote the number of k-bit strings which do *not* contain $t - 1$ consecutive zeros. Except for the string of all zeros (which we ignore), these are precisely the strings which do not contain the substring $v_t = \overbrace{10\cdots0}^{t}$. The proofs of Lemmas 3-1 through 3-4 depend heavily on the following combinatorial result.

Theorem 3-1: *For large t and k,*

$$\overline{\Psi}_k(t) \;=\; 2^k\, e^{-k2^{-t}}\, e^{O(t2^{-t},\,kt2^{-2t})}.$$

Proof: We first count the number $\overline{\Psi}_k{}'(t)$ of k-bit strings which do not contain an occurrence of v_t between the beginning and end of the string (i.e., for the time being we ignore the occurrences of v_t which begin at the end and end at the beginning of a string).

Fix t and let f_i denote the number of i-bit strings ending with v_t but which do not contain any other occurrences of v_t in the string. Set $F(x) = \sum_{i=0}^{\infty} f_i x^i$. Note that $\overline{\Psi}_k{}'(t)$ is the $(k+t)th$ coefficient of $F(x)$. Let $f_i^{(j)}$ denote the number of i-bit strings ending in v_t which contain precisely j occurrences of v_t and set

$$F^{(j)}(x) \;=\; \sum_{i=0}^{\infty} f_i^{(j)}\, x^i$$

Since occurrences of v_t cannot overlap, it is not difficult to show that $F^{(j)}(x)$ is identical to $F(x)^j$ for all $j > 1$

Let g_i be the number of i-bit strings which end in v_t (regardless of the number

of other occurrences of v_t which appear in the string) and set $G(x) = \sum_{i=0}^{\infty} g_i x^i$
Since $g_i = 2^{i-t}$ for all $i \geq t$, it is easily seen that $G(x) = x^t/(1-2x)$. Also note
that

$$G(x) = \sum_{j=1}^{\infty} F^{(j)}(x)$$

$$= \sum_{j=1}^{\infty} F(x)^j$$

$$= [1/(1 - F(x))] - 1$$

and thus that

$$F(x) = G(x)/(G(x) + 1)$$

$$= x^t/(1 - 2x + x^t)$$

Thus $\overline{\Psi}_k{}'(t)$ is simply the kth coefficient of $1/(1 - 2x + x^t)$. For example,
$\overline{\Psi}_4{}'(2) = 5$ which is the coefficient of x^4 in the expansion of $1/(1 - 2x + x^2)$.

Let $p(x) = 1 - 2x + x^t$. It is easily observed that $gcd(p(x), dp(x)/dx) = 1$
and thus that $p(x)$ does not have any multiple roots for $t > 2$. Thus we can
expand

$$p(x)^{-1} = \sum_{i=1}^{t} A_i/(x-r_i)$$

where $\{r_i | 1 \leq i \leq t\}$ is the set of distinct (and possibly complex) roots of $p(x)$ and

$$A_i = [(x-r_i)/p(x)]_{x=r_i}$$

$$= 1/[dp(x)/dx]_{x=r_i}$$

for $1 \leq i \leq t$. Once the roots of $p(x)$ are known, we can calculate $\overline{\Psi}_k{}'(t)$ from
the formula

$$\overline{\Psi}_k{}'(t) = -\sum_{i=1}^{t} A_i r_i^{-(k+1)}$$

Although we do not know how to find the roots of $p(x)$ explicitly for large t, we
can describe them asymptotically. First observe that as $t \to \infty$, the absolute value
of every root must approach either $1/2$ or 1. Otherwise the absolute value of one

term of $p(x)$ will dominate the sum of the absolute values of the other two terms. For example, if $|r| \leq c < 1/2$ as $t \to \infty$ for some root r and constant c, then $1 > |2r| + |r^t|$ for large t.

For any root r such that $|r| \to 1$, the absolute value of r must be greater than or equal to 1 for large t. Otherwise there would be a root r and a function $\varepsilon(t) \to 0^+$ such that $|r| = 1 - \varepsilon(t)$. But then

$$|2r| = 2 - 2\varepsilon(t)$$
$$> 1 + |1 - \varepsilon(t)|^t$$
$$= 1 + |r^t|$$

and it would be impossible for $p(r)$ to vanish for $t>2$, a contradiction.

For any root r such that $|r| \to 1/2$, it is necessary that $r \to 1/2$. Otherwise, the real part of $p(r)$ cannot vanish for large t. An even more precise description of such roots can be found by substituting $(1/2)e^{s(t)}$ for r where $s(t) \to 0$ as $t \to \infty$. This substitution yields

$$1 - e^{s(t)} + 2^{-t} e^{ts(t)} = 0$$

and thus

$$1 - (1 + s(t) + O(s(t)^2)) + 2^{-t}(1 + O(ts(t))) = 0.$$

Thus $s(t) = 2^{-t} + q(t)$ where $|q(t)| \ll 2^{-t}$ as $t \to \infty$. Another iteration of this process reveals that $q(t) = O(t2^{-2t})$ and thus that

$$r = (1/2) e^{2^{-t}} e^{O(t2^{-2t})} \quad \text{as } t \to \infty .$$

In fact, there is precisely one root, say r_1, which approaches $1/2$ as $t \to \infty$. All other roots approach 1 in absolute value. To prove this fact, we first note that at least one root approaches $1/2$ in absolute value. Otherwise, the absolute value of the product of all of the roots would be strictly larger than 1. This contradicts the fact that the constant term in $p(x)$ is exactly 1. On the other hand, if $p(x)$ had two or more roots near $1/2$, then the preceding expression for r coupled with the knowledge that the other roots are near the unit circle could be used to show that $|[dp(x)/dx]_{x=r_i}| < O((3/2)^{l-2}l2^{-2l}) = o(1)$. This contradicts the easily observed fact that $[dp(x)/dx]_{x=r_i} \approx -2$.

It remains to compute the A_i. Since $dp(x)/dx = lx^{l-1} - 2$ we find that $A_1 = -(1/2) + O(l2^{-l})$ and that $A_i = O(1/l)$ for $2 \le i \le l$ Thus

$$\overline{\Psi}_k{}'(l) = O(l) \cdot [-1/2 + O(l2^{-l})] 2^{k+1} e^{-(k+1)2^{-l}} e^{O(kl2^{-2l})}$$

Replacing $1 + O(l2^{-l})$ with $e^{O(l2^{-l})}$ and simplifying, we conclude that

$$\overline{\Psi}_k{}'(l) = 2^k e^{-k2^{-l}} e^{O(l2^{-l}, kl2^{-2l})}$$

for large l and k.

The only strings which are included in the count of $\overline{\Psi}_k{}'(l)$ but not in that of $\overline{\Psi}_k(l)$ are those of the form $\overbrace{0\cdots0}^{i}w\overbrace{10\cdots0}^{l-i}$ where $1 \le i \le l-1$ and w is a string which is included in the count of $\overline{\Psi}_{k-l}{}'(l)$. Thus

$$\overline{\Psi}_k(l) = \overline{\Psi}_k{}'(l) - (l-1)\overline{\Psi}_{k-l}{}'(l)$$
$$= 2^k e^{-k2^{-l}} e^{O(l2^{-l}, kl2^{-2l})} - (l-1) 2^{k-l} e^{-(k-l)2^{-l}} e^{O(l2^{-l}, kl2^{-2l})}$$
$$= 2^k e^{-k2^{-l}} e^{O(l2^{-l}, kl2^{-2l})}$$

for large l and k. This completes the proof of the theorem. $\square$

We can now prove Lemmas 3-1 and 3-2.

Proof of Lemma 3·1: From the definition, we know that

$$\Psi_k(l) = \overline{\Psi}_k(l+2) - \overline{\Psi}_k(l+1)$$
$$= 2^k e^{-k2^{-(l+2)}} e^{O(l2^{-l}, kl2^{-2l})} \quad 2^k e^{-k2^{-(l+1)}} e^{O(l2^{-l}, kl2^{-2l})}$$

for large l and k. For $l \ge (\log k)/2 + \log\log k$, both $l2^{-l}$ and $kl2^{-2l}$ vanish as $k \to \infty$. In what follows, we will show that if $l \ll k$ then

$$e^{-k2^{-(l+2)}} - e^{-k2^{-(l+1)}} \gg O(l2^{-l}, kl2^{-2l})$$

and thus that

$$\Psi_k(t) \sim 2^k (e^{-k2^{-(t+2)}} \quad e^{-k2^{-(t+1)}})$$

Assume for the purposes of contradiction that

$$e^{-k2^{-(t+2)}} \cdot e^{-k2^{-(t+1)}} \leq O(t2^{-t}, kt2^{-2t})$$

Then, $e^{-k2^{-(t+2)}} \sim e^{-k2^{-(t+1)}}$ which means that $e^{-k2^{-(t+2)} + k2^{-(t+1)}} \sim 1$ and thus that $k2^{-(t+2)} \to 0$ Thus we can use a Taylor series expansion of the exponentials to find that

$$e^{-k2^{-(t+2)}} \cdot e^{-k2^{-(t+1)}} \sim (1 - k2^{-(t+2)}) \quad (1 - k2^{-(t+1)})$$

$$= k2^{-(t+2)}$$

$$\gg O(t2^{-t}, kt2^{-2t})$$

provided that $t \ll k$ a contradiction. $\square$

Proof of Lemma 3-2: The number of k-bit strings which do not contain a block of $logk - loglnk - 1$ consecutive zeros is

$$\overline{\Psi}_k(logk - loglnk) \sim 2^k e^{-k2^{-logk + loglnk}}$$

$$= 2^k/k$$

$$= O(N/logN) .$$

The number of k-bit strings which contain a block of $2logk + 1$ consecutive zeros is

$$2^k \quad \overline{\Psi}_k(2logk+2) \sim 2^k - 2^k e^{-k2^{-2logk-2}} e^{O((logk)/k^2)}$$

$$= 2^k - 2^k[1 - 1/(4k) + O((logk)/k^2)]$$

$$\sim 2^k/4k$$

$$= O(N/logN) .\square$$

The proofs of Lemmas 3-3 and 3-4 depend on the following corollary to

Theorem 3-1.

Corollary 3·1: *For bounded m and p and large k and t,*

$$\sum_{t=1}^{(k+p)/m} \overline{\Psi}_{k-mt+p}(t) = O(2^k/k^m)$$

Proof: We first observe that for $t < 2logk/3$,

$$\overline{\Psi}_{k-mt+p}(t) \leq \overline{\Psi}_k(2logk/3)$$
$$\sim 2^k e^{-k2^{-(2logk)/3}}$$
$$= 2^k e^{-k^{1/3}}$$

and thus that

$$\sum_{t=1}^{(2logk)/3} \overline{\Psi}_{k-mt+p}(t) \leq (2/3) logk \, 2^k e^{-k^{1/3}}$$
$$\ll 2^k/k^m$$

for any finite m and p as $k \to \infty$

For larger values of t,

$$\overline{\Psi}_{k-mt+p}(t) \sim 2^{k-mt+p} e^{-k2^{-t}}$$

and thus

$$\sum_{t=(2logk)/3}^{(k+p)/m} \overline{\Psi}_{k-mt+p}(t) \sim \sum_{t=(2logk)/3}^{(k+p)/m} 2^{k-mt+p} e^{-k2^{-t}} .$$

By making the change of variables $r = t - logk$, we can see that the preceding sum is at most

$$(2^{k+p}/k^m) \sum_{r=-\infty}^{\infty} 2^{-mr} e^{-2^{-r}}$$

and thus at most $O(2^k/k^m) = O(N/logN)$. □

Proof of Lemma 3·3: A string whose longest block of zeros has length t and whose second longest block of zeros has length $s \leq t$ is of the form $w\overbrace{10\cdots0}^{t+1}w'$, where the longest block of zeros in ww' has length s. By definition, there are at most $k\Psi_{k-t-l}(s)$ such strings. Thus the sum over all *necklaces* of the difference between the sizes of the longest block and second longest block of zeros is at most

$$\leq (1/k) \sum_{t=0}^{k-l} \sum_{s=0}^{t} (t\text{-}s)\, k\, \Psi_{k-t-l}(s)$$

$$= \sum_{t=0}^{k-l} \sum_{s=0}^{t} (t\text{-}s)\, [\overline{\Psi}_{k-t-l}(s+2) - \overline{\Psi}_{k-t-l}(s+l)]$$

$$= \sum_{s=l}^{k} \sum_{t=s}^{k} \overline{\Psi}_{k-t}(s)$$

$$= \sum_{s=l}^{k} \left(2^k\, e^{-k2^{-s}}\, e^{O(s2^{-s},\, ks2^{-2s})} \sum_{t=s}^{k} 2^{-t}\, e^{t2^{-s}} \right)$$

$$\leq \sum_{s=l}^{k} \left(2^k\, e^{-k2^{-s}}\, e^{O(s2^{-s},\, ks2^{-2s})}\, 2^{-s}\, e^{O(s2^{-s})} \right)$$

$$= \sum_{s=l}^{k} 2^{k-s}\, e^{-k2^{-s}}\, e^{O(s2^{-s},\, ks2^{-2s})}$$

$$\leq \sum_{s=l}^{k} \overline{\Psi}_{k-s}(s)$$

$$= O(N/\log N)$$

by Corollary 3-1. □

Proof of Lemma 3-4:   Consider a necklace which fails to have a uniquely distinguished node. Each node in such a necklace must have one of the following three forms:

1) $w_l\overbrace{0\cdots0}^{l/2}w_2\underbrace{0\cdots0}w_3$,

2) $w_l\overbrace{0\cdots0}w_2\overbrace{0\cdots0}w_3\underbrace{0\cdots0}w_4$, or

3) $w_l\overbrace{0\cdots0}w_2\underbrace{0\cdots0}w_3\overbrace{0\cdots0}w_4\underbrace{0\cdots0}w_5$

where t is the length of the longest block of zeros in any of the strings. It is easily seen that there are at most

1) $k \sum_{t=1}^{N/2} \overline{\Psi}_{k-2t}(t+2)$ nodes of the first type,

2) $k^2 \sum_{t=1}^{N/3} \overline{\Psi}_{k-3t}(t+2)$ nodes of the second type, and

3) $k^3 \sum_{t=1}^{N/4} \overline{\Psi}_{k-4t}(t+2)$ nodes of the third type.

By Corollary 3-1, we can thus conclude that there are at most $O(N/logN)$ such nodes altogether. $\square$

CHAPTER 4

PRACTICAL LAYOUTS

Although the $O(N^2/log^2N)$-area layout for the shuffle-exchange graph described in Chapter 3 is (up to a constant) *asymptotically* optimal, it is not optimal for small values of N (e.g., $N=128$). In fact, none of the general layout procedures thus far discussed provide good layouts for small shuffle-exchange graphs. For practical applications, however, these are precisely the shuffle-exchange graphs for which we need good layouts.

In this chapter, we describe techniques for finding good layouts for small shuffle-exchange graphs. Although the techniques (which are described in Section 4.2) do not yet constitute a general procedure for finding truly optimal layouts for all shuffle-exchange graphs, they can be used to find "very nice" layouts for "small" shuffle-exchange graphs. As examples, we have included layouts for the *8*-node, *16*-node, *32*-node, *64*-node and *128*-node shuffle-exchange graphs in Section 4.3. The layouts are "very nice" in the sense that:

1) they require much less area than previously discovered layouts,

2) they have a certain natural structure which facilitates efficient layout description, chip manufacture and I/O management, and

3) they require the minimal amount of area for layouts with such structure.

4.1 Preliminaries

We have chosen to use the Thompson grid model [93] to illustrate our techniques because of its widespread acceptance and its simplicity. For practical layouts, however, the assumption that processors can be represented by points is clearly false. Nonetheless, we show in Section 4.1.1 that good Thompson model layouts can still be used to find good practical layouts. Thus we will be able to rest assured that the Thompson model is, in fact, an acceptable means for describing practical layouts of the shuffle-exchange graph.

We must also be sure that the *layouts* we design can be effectively used in practice. For example, it is important that the layouts have a suitable input/output structure so that data can be put on and taken off the chip efficiently. In Section 4.1.2, we describe a general class of layouts for the shuffle-exchange graph which appear to satisfy such constraints. The remainder of the chapter will then be devoted to finding optimal layouts within this class.

4.1.1 A Closer Look at the Thompson Model

The manner in which the Thompson model is useful for describing practical layouts varies with the size of the processors involved. For example, if one desires to use the shuffle-exchange graph as a permuter, then each processor need only contain k storage registers and some I/O hardware. Such a processor can be easily hardwired in a $k{\times}k$ square. In order to achieve maximum parallelism, each wire of the Thompson model layout is reproduced k times so that an entire k-bit word can be transmitted in one time step. For example, the optimal $2{\times}6$ Thompson model layout for the 8-node shuffle-exchange graph (which is shown in Figure 4-3 in Section 4.3) can be transformed into the more realistic $6{\times}18$ layout shown in Figure 4-1 by tripling the grid lines and replacing the point processors by $3{\times}3$ boxes (into which the guts of each processor can later be wired).

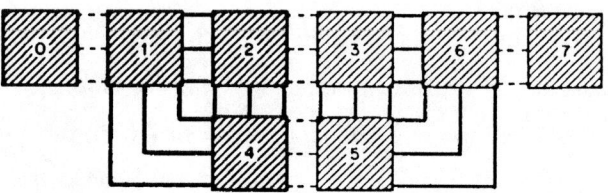

Figure 4·1: *A transformed Thompson model layout*
for the 8-node shuffle-exchange graph.

For some applications, the processors themselves require an entire chip. For example, every processor of a shuffle-exchange graph used to compute discrete Fourier transforms must be equipped with a floating point multiplier. Using the best technology currently available, only a few floating point multipliers can be wired onto a single chip. In this case, a Thompson model layout can be used to design an efficient *layout of chips* where each chip contains a single processor. The wires, as before, are replicated to achieve maximum parallelism but now serve as links between chips. Since the wires must be much wider in such a device, the side length of a processor (the chip) is about the same as the combined width of all the wires (pins) attached to it. By following an expansion procedure similar to the one described in the previous example, a good Thompson model layout can thus be used to design a good practical layout.

4.1.2   A Class of Practical Layouts

In this chapter, we will consider layouts for the shuffle-exchange graph for which:

  1) each necklace appears as a rectangle consisting of arbitrarily long segments of two vertical tracks and unit length segments of two

horizontal tracks,

2) the vertical tracks are divided into pairs, each pair containing at most one full necklace and any number of degenerate necklaces, and

3) each exchange edge appears as a horizontal line segment.

For example, the layouts described in Chapter 2 have this form.

Such layouts are particularly well suited for practical implementation since their structure facilitates efficient description, chip manufacture and data management. For example, by attaching a pin to each of the $O(N/logN)$ necklaces (this *is* feasible for small N), it is possible to load N input values into an N-processor shuffle-exchange chip in just $O(logN)$ steps.

Even more importantly, we will show in the following section how to find layouts with the above form which require very small amounts of area. Thus very little is lost by restricting our attention to such layouts.

4.2 Optimization Techniques

In this section, we explain how to find layouts for small shuffle-exchange graphs which are optimal up to the constraints described in Section 4.1.2. For the most part, our methods are comprised of common sense, heuristics and exhaustive searches.

4.2.1 Ordering the Necklaces

The first step in finding optimal layouts of the form described in Section 4.1.2 is to order the necklaces from left to right so that the number of exchange edges which overlap at each point of the ordering is kept small. More precisely, we wish to find an ordering of the necklaces for which the maximum number of exchange edges overlapping at any point is minimized. For example, no more than 6 exchange edges overlap at any point of the ordering used to produce the layout for the *32*-node shuffle-exchange graph shown in Figure 4-2. If we switched the necklace ⟨*5*⟩ with ⟨*11*⟩, however, 9 exchange edges would overlap in the gap between ⟨*7*⟩ and ⟨*5*⟩. Since the maximum overlap is a lower bound on the number of horizontal tracks necessary to insert the exchange edges, we can easily see that the latter ordering is inferior since any layout it produces must have at least 9 horizontal tracks. Note that the layout in Figure 4-2 has just 6 horizontal tracks.

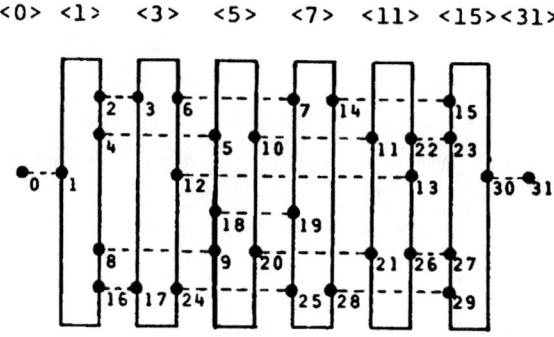

Figure 4-2: *A good ordering of the necklaces for the 32-node shuffle-exchange graph.*

As we mentioned in Chapter 3, it is not known how best to order the necklaces in general. For small shuffle-exchange graphs, however, there are several simple heuristics which produce optimal orderings. For example, arrangements of the necklaces from left to right in order of nondecreasing size and arrangements in order of increasing minimal number represented are usually quite close to optimal for small shuffle-exchange graphs. In fact, such orderings are within a necklace swap of optimal for $N \leq 256$ ($k \leq 8$). Note the the ordering displayed in Figure 4-2 could have been produced by either of these methods.

Probably the most **difficult task** is proving that a good ordering is, in fact, optimal. The techniques **we have** used to prove optimality depend heavily on exhaustive searches. For *k≤8*, the techniques have succeeded in proving the optimality of good orderings. **For** *9≤k≤13*, we have found good orderings but **have been unable to prove that they are optimal.** We have summarized the results **in Table 4-1. Note that for each** *k*, **the** maximum overlap of the best known ordering serves only as a *lower* bound for the number of horizontal tracks that will be required for any layout with that ordering. In some cases, additional horizontal tracks may be required.

Table 4-1

Maximum Overlap of Best Known Orderings

k	N	maximum overlap of best known ordering	optimal?
3	8	2	yes
4	16	3	yes
5	32	6	yes
6	64	10	yes
7	128	18	yes
8	256	33	yes
9	512	62	?
10	1024	115	?
11	2048	214	?
12	4096	388	?
13	8192	754	?

4.2.2 Inserting the Exchange Edges

The second step in constructing optimal layouts for small shuffle-exchange graphs is to insert the exchange edges using as few horizontal tracks as possible. Recall that in Chapter 2, we showed how to use the complex plane diagram as one method of inserting the exchange edges. Although this method is theoretically nice, it is not very practical since it uses an excessive number of horizontal tracks to insert the exchange edges. For example, *10* horizontal tracks were used to insert the exchange edges in the layout shown in Figure 2-3 whereas only *6* tracks were required in the layout shown in Figure 4-2 (even though the same necklace orderings were used for both layouts).

The complex plane diagram can still be of use when inserting exchange edges, however. For example, notice that the top-to-bottom orderings of the exchange edges across most of the vertical cuts which are located between necklaces in the layout in Figure 4-2 are the same as the orderings for the corresponding cuts in Figure 2-3. In general, knowledge of the level structure of the complex plane diagram is very helpful in optimizing the insertion of the exchange edges. In fact, we relied heavily on such knowledge when constructing the optimal layouts displayed in Section 4.3.

For very small shuffle-exchange graphs (e.g., for $k \leq 5$), it is possible to find optimal embeddings of the exchange edges by trying all reasonable possibilities. For somewhat larger shuffle-exchange graphs (e.g., $k=6,7$), however, the task is substantially more difficult. In order to find the optimal layouts shown in Section 4.3, we

1) first located the center of the region of maximum overlap and (using the complex plane diagram as a guide) inserted the exchange edges which crossed the region (one edge on each horizontal track),

2) next inserted the exchange edges located in neighboring regions without

(if possible) introducing any additional tracks, and

3) lastly inserted the remaining exchange edges (again without adding any new horizontal tracks).

Steps 1 and 3 are easy but step 2 can be difficult. In some cases it is necessary to interchange the left and right parts of some necklaces or to slide a node around from one part of a necklace to the other. For $k = 6$ and 7, it is also necessary to introduce an extra horizontal track at step 2. For larger shuffle-exchange graphs, it would probably be necessary to introduce even larger numbers of horizontal tracks.

4.2.3   Additional Savings

All of the practical layouts we have considered thus far have two horizontal tracks which are used solely for the purpose of connecting the left part of each necklace to the right part. It is not difficult to show that these tracks can be eliminated without affecting the rest of the layout. As an example of how this can be accomplished, we suggest that the reader compare the layout of the *32*-node shuffle-exchange graph shown in Figure 4-2 with that in Figure 4-5.

Even larger savings can be had for some shuffle-exchange graphs by doubling up the degenerate necklaces with full necklaces in the same pair of vertical tracks, thus reducing the number of vertical tracks used. Of course, it is necessary to rearrange the exchange edges somewhat but, as degenerate necklaces have very few nodes in small shuffle-exchange graphs, this can usually be done without introducing any additional horizontal tracks. For example, substantial savings can be achieved in this manner for the *16*-node and *64*-node shuffle-exchange graphs.

4.3 Optimal Layouts

In the following figures, we exhibit layouts for the *8*-node, *16*-node, *32*-node, *64*-node and *128*-node shuffle-exchange graphs which are optimal up to the constraints described in Section 4.1.2. The layouts were found via the techniques described in Section 4.2.

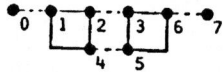

Figure 4-3: *A 2x6 layout for the 8-node shuffle-exchange graph.*

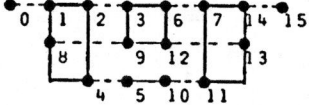

Figure 4-4: *A 3x8 layout for the 16-node shuffle-exchange graph.*

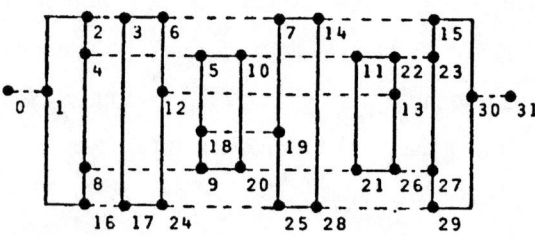

Figure 4-5: *A 6x14 layout for the 32-node shuffle-exchange graph.*

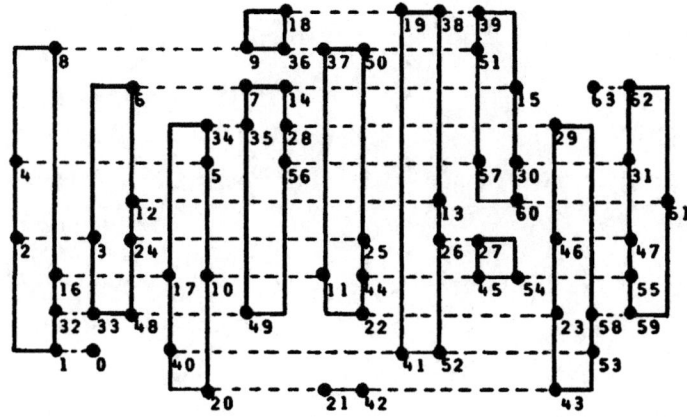

Figure 4-6: *An 11x18 layout for the 64-node shuffle-exchange graph.*

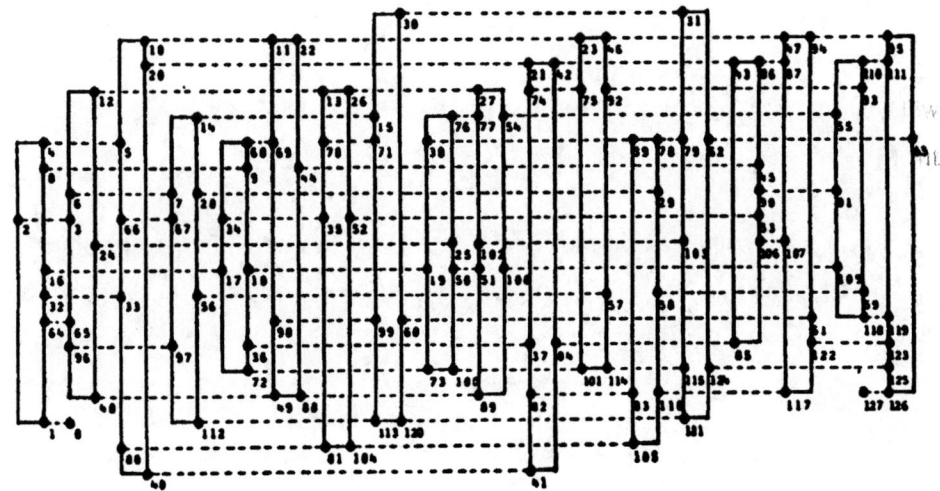

Figure 4-7: *A 19x36 layout for the 128-node shuffle-exchange graph.*

4.4   Other Layouts

To this point, we have considered only a specific class of layouts for the shuffle-exchange graph. As these layouts are quite good, it is not clear that we need to consider others. Nevertheless, it is worth pointing out that slightly better layouts do exist for some shuffle-exchange graphs. For example, by considering layouts in which the exchange edges are allowed to bend and in which two or more full necklaces can occupy the same pair of vertical tracks, it is possible to construct the layout for the *32*-node shuffle-exchange graph shown in Figure 4-8.

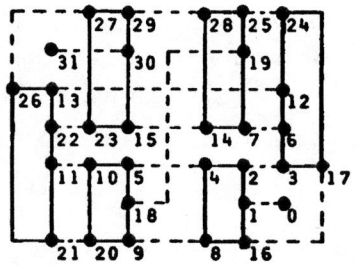

Figure 4-8: *An improved 7x9 layout for the 32-node shuffle-exchange graph.*

It is likely that slight improvements can also be made for larger shuffle-exchange graphs. At this point, however, we feel that research efforts should be directed more towards implementation of the good layouts already discovered. Once this is done, it will be much clearer whether or not the effort necessary to further reduce the layout area is justified.

PART II

LOWER BOUND TECHNIQUES

CHAPTER 5

REVIEW OF KNOWN TECHNIQUES

In this chapter, we review the known techniques for determining the layout area, crossing number and maximum edge length of an arbitrary VLSI network. We also preview some of the results that will be proved in Chapters 6 through 8. A comparison of the new lower bounds with the previously known upper and lower bounds can be found in Tables 5-2, 5-3 and 5-5.

5.1 Area Bounds

One of the most important problems in VLSI layout theory is the determination of the minimum amount of area required to lay out a network on a chip. Given an arbitrary graph, this problem has two parts; namely,

1) finding a good layout for the graph, and

2) showing that the layout is optimal.

There are a variety of techniques known for finding good layouts for specific graphs [3, 15, 26, 27, 34, 40, 47, 51, 52, 55, 62, 65, 67, 74, 75, 84, 89, 93, 94, 97, 98, 100], but the only known general technique is due to Leiserson [55] and Valiant [98]. They showed how to construct a layout for any graph for which a separator is known. (An N-node graph is said to have an $f(x)$-*separator* if it can be partitioned into two equal-sized subgraphs G_1 and G_2 such that at most $f(N)$ edges link G_1 to

G_2 and both G_1 and G_2 have $f(x)$-separators. Although the notation is abusive, we will follow the common practice of writing $f(N)$ instead of $f(x)$ where N denotes the number of nodes in the graph.) We have summarized their results in Table 5-1.

Table 5-1

Upper Bounds on the Layout Area of
N-Node Graphs With Specified Separators

separator	upper bound on layout area
$O(N^\alpha)$, $\alpha < 1/2$	$O(N)$
$O(N^\alpha)$, $\alpha = 1/2$	$O(N log^2 N)$
$O(N^\alpha)$, $\alpha > 1/2$	$O(N^{2\alpha})$

There are two difficulties with Leiserson-Valiant method. First, it is not always possible to find a good separator for a graph. For instance, it is still not known whether or not the N-node shuffle-exchange graph has an $O(N/logN)$-separator. Second, the layouts produced by the technique are not always optimal — even if a minimal separator is known. For example, the technique requires $\Theta(N log^2 N)$ area to lay out the N-node mesh, substantially more than is really needed. For the most part, however, the method is a good one and certainly the most general technique currently available.

Once a good layout for a network has been found, it remains to show that the layout is optimal. This is accomplished by proving a good *lower* bound on the layout area of the network. The only known methods for proving such lower bounds are due to Thompson [92], Vuillemin [99] and Lipton and Sedgewick [59].

They have concentrated on the related problem of proving lower bounds for the bisection width of a graph. (The *bisection width* of a graph is the minimum number of edges which must be removed in order to separate the graph into two disjoint and equal-sized subgraphs.)

Thompson was the first to notice the relationship between bisection width and layout area. In particular, he showed that the *wire area* of a graph with bisection width B is at least $\Omega(B^2)$. In what follows, we prove the slightly weaker (and simpler) result for layout area.

Theorem 5-1 (Thompson [93]): *The layout area of a graph with bisection width B is at least $\Omega(B^2)$.*

Proof: Consider an optimal layout of a graph G with bisection width B. Cut the layout horizontally so that precisely $1/2$ of the nodes of G are above the cut. (For an example, see the diagram in Figure 5-1). Since at least B edges must cross the cut, the layout must contain at least B-1 vertical tracks. A similar argument reveals that the layout must also have at least B-1 horizontal tracks. Thus the area of the layout is at least $(B$-$1)^2 = \Omega(B^2)$. $\square$

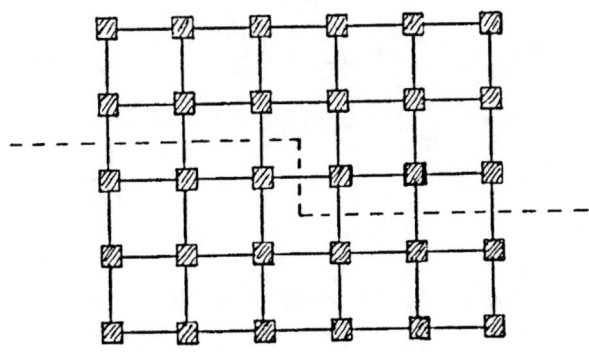

Figure 5-1: *A horizontal bisection of a layout.*

Although the task of finding a good lower bound on the bisection width of a graph is difficult in general, Thompson [93] was succesful in finding good bisection width lower bounds for a variety of computationally useful networks.   For example, he used information transfer arguments to show that any network which is capable of computing the discrete Fourier transform on N elements in T steps must have bisection width at least $\Omega(N/T)$.   Among other things, he was thus able to conclude that at least $\Omega(N^2/log^2N)$ area is required to lay out the N-node shuffle-exchange graph.

Thompson's work has recently been extended; first by Vuillemin [99] and then by Lipton and Sedgewick [59]. Vuillemin characterized a broad class of graphs for which Thompson's lower bound arguments can be applied while Lipton and Sedgewick showed how to use crossing sequence arguments to prove lower bounds for an even larger class of graphs.

Although the methods of Thompson, Vuillemin, Lipton and Sedgewick are quite elegant and useful in establishing good bisection width lower bounds for certain graphs, their applicability is inherently limited to graphs for which the layout area is no more than a constant times as large as the square of the bisection width. Thus they have not been of use in resolving two of the key open questions in VLSI layout theory; namely,

1) "How much area is needed to lay out a planar graph?" and

2) "How much area is needed to lay out a graph which has an $O(N^{1/2})$-separator?."

The planar graph question is particularly important since, as we will show in Chapter 7, the layout problem of an arbitrary graph can be reduced to that for a planar graph. No nontrivial lower bounds have been found for either problem, however.   As we mentioned previously, the best procedure known requires $O(Nlog^2N)$ area to lay out an arbitrary N-node graph with an $O(N^{1/2})$-separator.

As Lipton and Tarjan [60] have shown that every N-node planar graph has an $O(N^{1/2})$-separator, the $O(N\log^2 N)$-area layout procedure also works for planar graphs. Although it is suspected that better layout procedures exist for planar graphs, none have yet been found.

In Part II, we pursue an entirely different strategy in developing new lower bound techniques for VLSI. Whereas previous researchers have been concerned primarily with the bisection width of a network, we shall be concerned with its *crossing number* and *wire area*. Both are lower bounds on the layout area of any graph. In fact, we will show in Chapter 7 that

$$\Omega(B^2) \leq C+N \leq W \leq O(A)$$

for any N-node graph with bisection width B, crossing number C, wire area W and layout area A.

The preceding inequality implies that every lower bound technique for the bisection width of a graph is also a lower bound technique for its crossing number and wire area. Thus nothing is lost by forgetting about bisection width and concentrating ones efforts on finding good lower bounds for the crossing number and wire area of a graph. In fact, much can be gained. For example, we will use such techniques to find

1) an N-node planar graph which has layout area $O(N\log N)$, and

2) an N-node (nonplanar) graph with an $O(N^{1/2})$-separator which has layout area $O(N\log^2 N)$.

The first result demonstrates that not all planar graphs can be laid out in linear area, thus disproving a conjecture thought by many to be true. The second result indicates that the Leiserson-Valiant $O(N\log^2 N)$-area layout technique for graphs with $O(N^{1/2})$-separators is optimal at least some of the time and thus cannot, in general, be improved.

For easy reference, we have summarized our results along with the previously known upper and lower bounds in the following table. The upper bounds are from Table 5-1 and represent the maximal amount of area needed to lay out any graph with the designated property. The lower bounds, on the other hand, represent the minimal amount of area required to lay out a *specific* class of graphs with the designated property. The previously known lower bounds are, for the most part, trivial. The only exception is the $N^{2\alpha}$ bound which, as a corollary of Theorem 5-1, is due to Thompson [93].

Table 5-2

Area Bounds

separator	previous lower bound	our lower bound	upper bound
$O(N^\alpha)$, $\alpha < 1/2$	$\Omega(N)$		$O(N)$
$O(N^\alpha)$, $\alpha = 1/2$	$\Omega(N)$	$\Omega(Nlog^2N)$	$O(Nlog^2N)$
$O(N^\alpha)$, $\alpha > 1/2$	$\Omega(N^{2\alpha})$		$O(N^{2\alpha})$
(planar)	$\Omega(N)$	$\Omega(NlogN)$	$O(Nlog^2N)$

5.2 Crossing Number Bounds

Surprisingly, the crossing number problem has not been previously studied in the context of VLSI. Hence no nontrivial bounds on the number of wire crossings have been developed, even for networks like the shuffle-exchange graph.

In Chapter 7, we study the crossing number problem in detail and develop a variety of techniques for proving both upper and lower bounds. We first prove the inequality $C + N \geq \Omega(B^2)$ mentioned in Section 5.1. As a corollary, we then generalize Theorem 5-1 to prove crossing-time tradeoffs for chips that compute transitive functions. As an application of the tradeoff, we show that the crossing number of the N-node shuffle-exchange graph is $\Theta(N^2/log^2N)$. We also develop more sophisticated and powerful methods for proving crossing number bounds.

A summary of the new bounds is included in Table 5-3. The previous upper bounds follow trivially from the area upper bounds in Table 5-1. As before, the upper bounds are universal while the lower bounds are existential in nature.

Table 5-3

Crossing Number Bounds

separator	our lower bound	our upper bound	previous upper bound
$O(N^\alpha)$, $\alpha < 1/2$	$\Omega(N)$		$O(N)$
$O(N^\alpha)$, $\alpha = 1/2$	$\Omega(NlogN)$	$\Omega(NlogN)$	$O(Nlog^2N)$
$O(N^\alpha)$, $\alpha > 1/2$	$\Omega(N^{2\alpha})$		$O(N^{2\alpha})$

The lower bound for graphs with $O(N^\alpha)$-separators where $\alpha<1/2$ is straightforward. The example is a collection of $N/5$ disjoint 5-cliques. This graph clearly has an $O(1)$-separator and at least $N/5$ wire crossings. The lower bound for the case when $\alpha>1/2$ is also easy given the fact (which we prove in Chapter 7) that $N+C \geq \Omega(B^2)$ for any graph. The bounds when $\alpha=1/2$ are substantially more difficult. In the case of planar graphs, there is a tradeoff between crossing number and area. For example, Valiant [98] has discovered N-node graphs that have $O(N)$-area layouts with $O(N)$ crossings but for which any planar layout has $O(N^2)$ area.

5.3   Edge Length Bounds

There has been a great deal of interest lately in the problem of minimizing the length of the longest wire in VLSI layouts [9, 12, 20, 26, 67, 84]. It is not difficult to show that the length of the longest wire in any reasonable, area-optimal VLSI layout is at most a constant times the square root of the layout area. (Otherwise, some wire would be longer than the perimeter of the layout, which is unreasonable.) Bhatt and Leiserson [9] recently found better layouts for graphs with small separators. We have summarized their results in Table 5-4. (It is worth noting that the layouts which achieve the bounds in Table 5-4 simultaneously achieve the best known bounds for layout area. Thus no *layout area/maximum edge length* tradeoffs are apparent.)

Table 5-4

Upper Bounds on the Maximum Edge Length of
N-Node Graphs With Specified Separators

separator	upper bound on maximum edge length
$O(N^\alpha)$, $\alpha < 1/2$	$O(N^{1/2}/logN)$
$O(N^\alpha)$, $\alpha = 1/2$	$O(N^{1/2}logN/loglogN)$
$O(N^\alpha)$, $\alpha > 1/2$	$O(N^\alpha)$

Very little has been accomplished in the way of lower bounds, however, since bisection width arguments do not seem to be applicable to edge length bounds. In fact, the only known lower bound for maximum edge length is the trivial lower bound derived from the diameter of a graph. (The *diameter* of a graph is the greatest distance between any pair of nodes in the graph where *distance* is defined to be the length of the shortest path linking the pair of nodes.) The precise lower bound is stated in the following theorem.

Theorem 5-2: *Any layout of a graph G with diameter d and minimum layout area A has some edge of length at least $A^{1/2}/3d$.*

Proof: Let Γ be any layout of G and q be the length of the longest wire in Γ. We will use Γ to construct another layout Γ' of G which has at most $9d^2q^2$ area. Since any layout for G has at least A area, this will be sufficient to show that $q \geq A^{1/2}/3d$.

Since every pair of nodes in G is linked by a path of length d or less, we can conclude that every pair of nodes are within distance dq of each other in Γ. (Otherwise, some edge would have length greater than q in Γ, a contradiction.) Thus, all of the nodes are contained in some $dq \times dq$ square in Γ. Since every wire which leaves the square must re-enter at some other point, we can conclude that at most $2dq$ wires can cross the boundary of the square. By rewiring the portion of Γ which is outside the square, it is possible to produce a second layout Γ' for G which has at most $2dq$ additional horizontal tracks and $2dq$ additional vertical tracks. (One additional horizontal track and one additional vertical track are needed to replace each wire.) Thus the total area of Γ' is at most $9d^2q^2$. (As an example of how the rewiring should be done, we have included Figure 5-2.) $\square$

It is not difficult to construct N-node graphs with $f(N)$-separators which have $logN$ diameter for any $f(N)$. By Theorem 5-2, any layout of such a graph must have a wire of length $\Omega(f(N)/logN)$. Using crossing number and wire area arguments, however, we will find examples of graphs which must contain even longer wires. In particular, we will describe

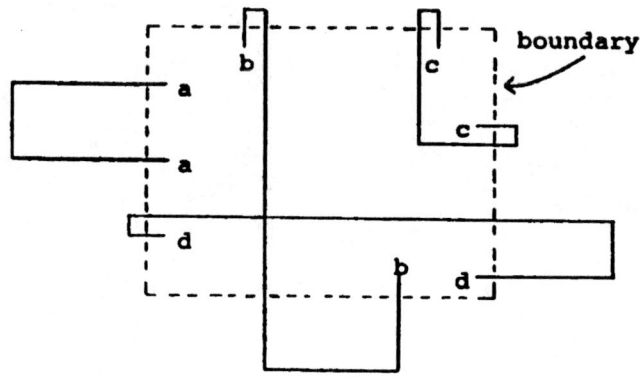

Figure 5-2: *Rewiring the outer portion of a layout.*

1) an N-node planar graph for which any layout must have a wire of length $\Theta(N^{1/2}/log^{1/2}N)$,

2) an N-node graph with an $O(N^{1/2})$-separator for which any layout must have a wire of length $\Theta(N^{1/2}logN/loglogN)$, and

3) an N-node graph with an $O(N^{1-1/r})$-separator for which any layout must have a wire of length $\Theta(N^{1-1/r})$ for any $r \geq 3$.

The latter two results achieve the known upper bounds for maximum wire length. They also indicate that some wires in some layouts must be very long (possibly as long as the length of the entire layout).

For convenience, we have summarized our edge length results along with the previously known upper and lower bounds in Table 5-5. The upper bounds are due to Bhatt and Leiserson [9] while the previously known lower bounds are all easy corollaries of Theorem 5-2.

Table 5-5

Maximum Edge Length Bounds

separator	previous lower bound	our lower bound	upper bound
$O(N^\alpha)$, $\alpha < 1/2$	$\Omega(N^{1/2}/logN)$		$O(N^{1/2}/logN)$
$O(N^\alpha)$, $\alpha = 1/2$	$\Omega(N^{1/2}/logN)$	$\Omega(N^{1/2}logN/loglogN)$	$O(N^{1/2}logN/loglogN)$
$O(N^\alpha)$, $\alpha > 1/2$	$\Omega(N^\alpha/logN)$	$\Omega(N^\alpha)$	$O(N^\alpha)$
(planar)	$\Omega(N^{1/2}/logN)$	$\Omega(N^{1/2}/log^{1/2}N)$	$O(N^{1/2}logN/loglogN)$

CHAPTER 6

NETWORK CONSTRUCTIONS

In this chapter, we describe the networks for which we will later establish layout area, crossing number and maximum edge length lower bounds. As the networks are new and interesting in their own right, we will discuss each at some length.

6.1 The 2-Dimensional Mesh of Trees

The N-node 2-dimensional mesh of trees will be the first example of a graph with an $O(N^{1/2})$-separator known to have layout area $O(N log^2 N)$, crossing number $O(N log N)$ and maximum edge length $O(N^{1/2} log N / log log N)$.

6.1.1. Definition

The *2-dimensional nxn mesh of trees* $M_{2,n}$ (where n is assumed to be a power of 2) is defined as follows. Starting with an *nxn* matrix of nodes and adding nodes wherever necessary, construct a complete binary tree in each row and column of the matrix. The trees should be constructed so that

1) the leaves in each tree are precisely the nodes in the corresponding row or column of the original matrix, and

2) the subgraph induced on the nodes in each quadrant is $M_{2,n/2}$.

For example, we have drawn $M_{2,4}$ in Figure 6-1. The nodes in the original *4x4*
matrix are represented by dots. The nodes which were added in order to form row
trees are drawn as small triangles while those added to form column trees are
shown as small squares. The row tree edges are drawn with solid lines while
dashed lines represent edges of column trees. Notice that if we were to remove the
roots of the row and column trees of $M_{2,4}$ and the edges incident to them, we
would be left with *4* copies of $M_{2,2}$, one in each quadrant. In general, if we
remove the nodes and edges in the top k levels of the binary trees in $M_{2,n}$, we
will be left with 2^{2k} copies of $M_{2,n2^{-k}}$. This important property of meshes of trees
is used extensively throughout Chapters 7 and 8.

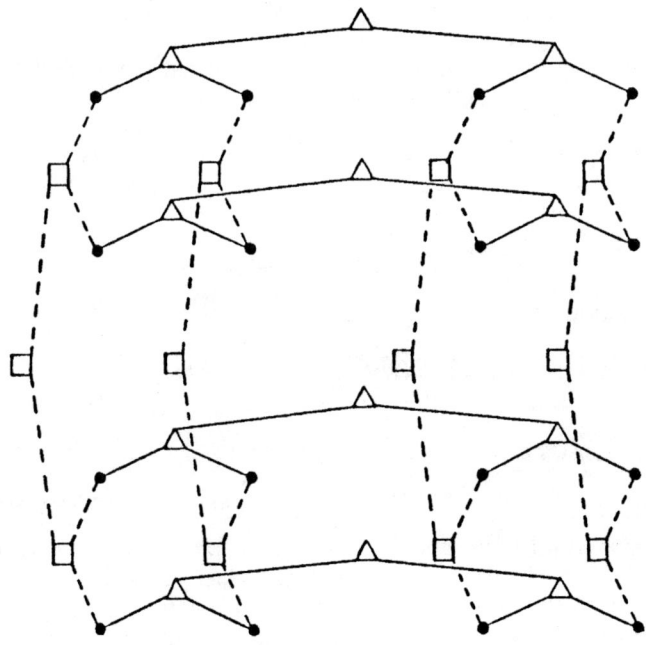

Figure 6-1: *The 4x4 mesh of trees $M_{2,4}$.*

6.1.2. Properties

It is not difficult to show that the $n \times n$ mesh of trees $M_{2,n}$ has

  1) $N = 3n^2 - 2n = \Theta(n^2)$ nodes,

  2) bisection width $n = \Theta(N^{1/2})$

  3) diameter $4logn = \Theta(logN)$ and

  4) an $O(N^{1/2})$-separator

By applying the methods discussed in Chapter 5, we can thus conclude that the N-node 2-dimensional mesh of trees has

  1) layout area between $\Omega(N)$ and $O(Nlog^2N)$,

  2) crossing number at most $O(Nlog^2N)$, and

  3) maximum edge length between $\Omega(N^{1/2}/logN)$ and $O(N^{1/2}logN/loglogN)$.

In fact, we will show in Chapters 7 and 8 that the N-node 2-dimensional mesh of trees has

  1) layout area $\Theta(Nlog^2N)$,

  2) crossing number $\Theta(NlogN)$, and

  3) maximum edge length $\Theta(N^{1/2}logN/loglogN)$.

Thus the 2-dimensional mesh of trees is the first graph with an $O(N^{1/2})$-separator known to acheive the upper bound for layout area discovered by Leiserson and Valiant and the upper bound for maximum edge length discovered by Bhatt and Leiserson.

6.1.3 Applications

Computationally, the $n \times n$ mesh of trees is a very powerful network. Among other things, it can be used to

1) multiply a fixed $n \times n$ matrix by m different n-vectors in $m + 2logn$ (word) steps,

2) sort a list of n m-bit words in $2m + 5logn$ (bit) steps, and

3) link n input terminals to n output terminals in any order in $logn$ (bit) steps.

The algorithms and processors needed for these operations are quite simple. For example, the processors needed for sorting and switching need only contain a few *and* and *or* gates while those for matrix-vector multiplication need only contain a word multiplier or adder. We describe the algorithms needed for these operations in the following three subsections.

(a) Matrix-Vector Multiplication

Given any fixed $n \times n$ matrix $S = (s_{ij})$, we will show how to program $M_{2,n}$ to compute the product of S and any m input n-vectors in $m + 2logn$ (word) steps. As S is fixed, it is not considered to be part of the on-line input. Rather, it is considered to be part of the program (in the form of off-line input) and thus we assume that the value of s_{ij} is initially stored in the (i,j) leaf of $M_{2,n}$ for each i and j. The algorithm proceeds as follows.

Given any input vector $v = (v_j)$, input the jth entry v_j into the root of the jth column tree for each j, $1 \leq j \leq n$. Pass the entries of v down the column trees so that after $logn$ steps, each leaf in the jth column tree has received the value of v_j. Computation of the n^2 products $\{s_{ij}v_j \mid 1 \leq i, j \leq n\}$ can now take place simultaneously. Afterwards, we can find the entries of the product vector Sv by summing the values of the leaves in each row tree. This operation takes an additional $logn$ steps.

The total running time of the algorithm just described is $1 + 2logn$. By pipelining the input vectors through the column trees and the output sums through the row trees, it is not difficult to see that m such products can be calculated in $m + 2logn$ steps.

(b)   Sorting

The algorithm for sorting proceeds as follows. Starting at the roots, input (bit by bit) the ith word to be sorted into the ith row and column trees for each i, $1 \leq i \leq n$. Pass the bits down each tree so that after $logn$ steps the leading bit of the ith word has reached each leaf of the ith row and column trees. Comparison of the ith and jth words for all i and j can now proceed simultaneously. After at most m additional steps, the (i,j) leaf has decided whether the ith word is smaller or larger than the jth word. Ties are broken arbitrarily (e.g., depending on the values of i and j). Once this is done, each leaf transmits a 0 or a 1 to its column tree father depending on whether its column tree word was smaller or larger than its row tree word. Each column tree then sums these values in order to determine the position of its word in the final ordering. (If the sum is carried out bit by bit starting with the least significant bit, this process takes $2logn$ steps.) This information is then used to mark a path in each column tree from the root to that leaf which is also in the appropriate row tree (again taking $2logn$ steps). It is now a simple matter to transmit the bits of the ith word along the unique path from the ith column tree root to the appropriate row root for each i. As the paths are all pairwise disjoint, this process takes only $m + 2logn$ steps.

The algorithm just described sorts a list of n m-bit numbers in $2m + 7logn$ steps. It is a simple exercise to speed up the alogorithm to obtain the $2m + 5logn$ step bound. We should also point out that this algorithm is similar to the one described by Muller and Preparata in [63]. The VLSI implementation of the algorithm is new, however, and far superior to many of the VLSI sorting algorithms discussed by Thompson in his recent survey paper [94].

(c) Switching

Given the algorithm just described for sorting, it is clear how to program $M_{2,n}$ to serve as a switching network for n input and output lines. For example, assume that the ith input line is to be connected to the jth output line for some i and j. In order to do this, we first hook up the ith input line to the ith column root. We next establish a path from the root of the ith cloumn tree to that leaf in the tree which is also in the jth row tree. This can be done by inspection of the binary representation $b_1 \cdots b_{logn}$ of the number j. More precisely, at the kth level of the binary tree, we branch left or right depending on whether b_k is 0 or 1 (respectively). Lastly, we link the appropriate leaf of the jth row tree to the root of the jth row tree and then to the jth output line (again taking $logn$ steps).

The algorithm just described takes $2logn$ steps to link n input lines to n output lines in any order. It is not difficult to show that if the row tree connections are hardwired in advance (i.e., by linking the root of each row tree to all of its leaves), then the input-output connections can be properly made in just $logn$ steps.

6.2   The r · Dimensional Mesh of Trees

The N-node r - dimensional mesh of trees (for $r > 2$) will be the first example of a graph with an $O(N^\alpha)$-separator (for $\alpha > 1/2$) known to have maximum edge length $\Theta(N^\alpha)$.

6.2.1   Definition

The 2-dimensional mesh of trees can be easily generalized to higher dimensions. For example, the *3-dimensional nxnxn mesh of trees* $M_{3,n}$ can be constructed as follows. Starting with an $nxnxn$ cube of nodes and adding nodes wherever necessary, construct a set of n^2 complete binary trees in each of the three dimensions of the cube. As before, the trees should be constructed so that the leaves are precisely the nodes of the original cube and so that the subgraph induced on each octant of nodes is $M_{3,n/2}$. The general r - *dimensional mesh of trees* $M_{r,n}$ is formed from an $\overbrace{nxnx \cdot \cdot \cdot xn}^{r}$ hypercube in a similar manner. In general, removal of the roots and edges which are in the top level of the binary trees will leave 2^r copies of $M_{r,n/2}$.

6.2.2   Properties

It is easily observed that the r - dimensional $\overbrace{nxnx \dots xn}^{r}$ mesh of trees $M_{r,n}$ has (for bounded r)

1) $N = (r+1)n^r - rn^{r-1} = \Theta(n^r)$ nodes,

2) bisection width $n^{r-1} = \Theta(N^{1-1/r})$,

3) diameter $2r\log n = \Theta(\log N)$,   and

4) an $O(N^{1-1/r})$-separator.

Thus we can easily infer that the N-node r - dimensional mesh of trees has (for bounded r)

1) layout area $\Theta(N^{2-2/r})$,

2) crossing number at most $O(N^{2-2/r})$, and

3) maximum edge length between $\Omega(N^{1-1/r}/\log N)$ and $O(N^{1-1/r})$.

In fact, we will show in Chapter 7 that the graph has

1) crossing number $O(N^{2-2/r})$, and

2) maximum edge length $\Theta(N^{1-1/r})$.

Thus the r-dimensional mesh of trees is the first graph with an $O(N^\alpha)$-separator (for $\alpha>1/2$) known to achieve the trivial upper bound on maximum edge length.

6.2.3   Application to Matrix Multiplication

Computationally, the r-dimensional mesh of trees is a very powerful network. For example, $M_{3,n}$ can be used to multiply m pairs of $n \times n$ matrices in $m + 2\log n$ (word) steps. The algorithm is very similar to the one used by $M_{2,n}$ to compute matrix-vector products.   It proceeds as follows.

At each time step, a pair of matrices is entered into the network via the roots of the trees in two of the dimensions (one dimension for each matrix).   The entries are passed down through the trees so that after $\log n$ steps, the leaf in the (r,s,t) position of the cube contains the (r,s) entry of the first matrix and the (s,t) entry of the second matrix for each r,s and t. All n^3 multiplications can then be performed simultaneously. The entries of the product matrix are then calculated by summing the values of the leaves of each tree in the third (previously unused) dimension. This process takes an additional $\log n$ steps.   As the network is easily pipelined, it is clear that the total computation time is just $m + 2\log n$ (word) steps.

6.2.4   A Further Generalization

The r-dimensional mesh of trees was defined as a natural generalization of the computationally powerful 2-dimensional mesh of trees. $M_{r,n}$ can also be viewed as a generalization of the r-cube, also a very powerful communications network. For example, $M_{r,2}$ is an r-cube with every edge replaced by a path of length 2. Viewed in this light, the r-dimensional mesh of trees motivates the definition of a *shuffle-tree graph* in the same way that the r-cube motivates the definition of the shuffle-exchange graph. Although we have yet to investigate this graph in detail, it is quite possible that it has important applications.

(As an aside, we should caution the reader that the asymptotic estimates given in Section 6.2.2 do not necessarily apply to $M_{r,2}$ since r was assumed to be bounded.   The correct estimates are not difficult to work out, however.)

6.3   The Tree of Meshes

The N-node tree of meshes will be the first example of a planar graph known to have $\Theta(N log N)$ layout area.

6.3.1   Definition

The *tree of meshes* is similar to the 2-dimensional mesh of trees in that it combines the structure of a mesh with that of a complete binary tree in a natural way. Unlike the 2-dimensional mesh of trees, however, the tree of meshes is a planar graph. It is formed by replacing each node of a complete binary tree with a mesh and each edge by several edges which link the meshes together. More precisely, the root of the binary tree is replaced by an $n x n$ mesh (where n is assumed to be a power of 2), its sons are replaced by $n/2$ x n meshes, their sons are replaced by $n/2$ x $n/2$ meshes, and so on until the leaves are replaced by $1 x 1$ meshes. In the place of each *right edge* of the binary tree (i.e., one which links a node to its right son), we link the rightmost column of nodes in the mesh corresponding to the father to the topmost row of nodes in the mesh corresponding to the right son. Similar replacements are made for *left edges* of the binary tree. In both cases, the connections are made so as to preserve the column and row order of the nodes and to insure that the resulting graph will be planar. The resulting graph is refered to as the $n x n$ tree of meshes and will be denoted by T_n . For example, we have drawn T_4 in Figure 6-2.

6.3.2   Properties

It is easily seen that the $n x n$ tree of meshes T_n has

1) $N = 2n^2 log n + n^2 = O(n^2 log n)$ nodes,

2) bisection width $n = O(N^{1/2}/log^{1/2}N)$,

3) diameter $8n = O(N^{1/2}/log^{1/2}N)$,   and

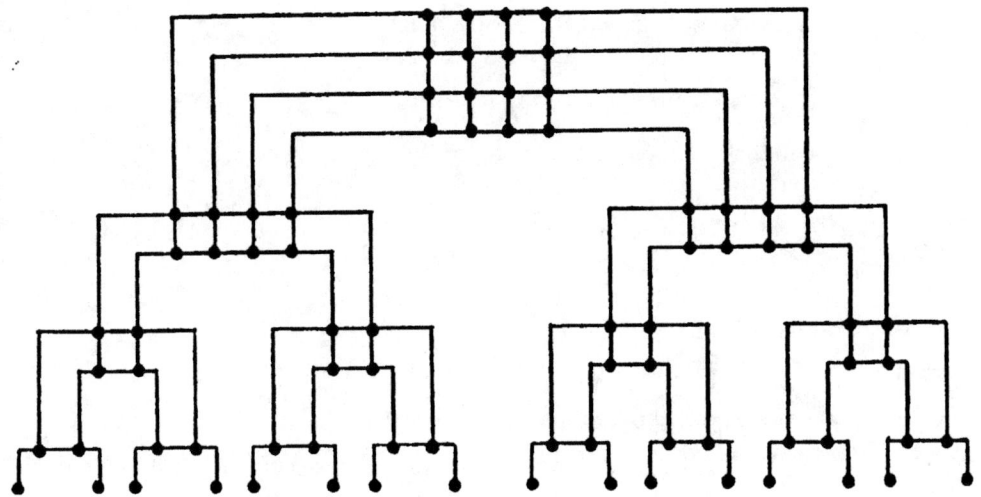

Figure 6-2: *The 4x4 tree of meshes T_4.*

4) an $O(N^{1/2})$-separator.

Thus we can easily infer that the N-node tree of meshes has

    1) layout area between $\Omega(N)$ and $O(N\log^2 N)$, and

    2) maximum edge length between $\Omega(\log^{1/2} N)$ and $O(N^{1/2}\log N/\log\log N)$.

In fact, the graph has

    1) layout area $\Theta(N\log N)$ and

    2) maximum edge length $O(\log N)$.

Both bounds are achieved by a straightforward modification of the *H*-tree layout [62] for a complete binary tree. (See [8] for a more detailed description of such layouts.) The layout area lower bound follows from Theorem 8-3, which states that the wire area of the N-node tree of meshes is $\Omega(N log N)$. Since the graph has $\Theta(N)$ wires, it is also true that some edge must have length $\Omega(log N)$, thus proving the maximum edge length lower bound.

In Section 6.4, we will show how to augment the N-node tree of meshes so that any layout will have to contain a wire of length at least $\Omega(N^{1/2}/log^{1/2}N)$.

6.3.3   Applications

The tree of meshes is a particularly interesting planar graph since it can embed arbitrary planar graphs much more efficiently than can the ordinary mesh. For example, it is not known how to embed an arbitrary planar graph in less than an $\Theta(N log^2 N)$-node mesh. As we show in part (a) of this section, however, any N-node planar graph can be embedded in an $\Theta(N log N)$-node tree of meshes.

The tree of meshes can also be used to embed many nonplanar graphs which have $\Theta(N^{1/2})$-separators. For example, we will show in part (b) of this section how to embed $M_{2,n}$ in T_{2n} for any n. This result will later allow us to give a simple proof that the N-node tree of meshes has wire area at least $\Omega(N log N)$.

(a)   Embeddings of Planar Graphs

In [60], Lipton and Tarjan prove an $\Theta(N^{1/2})$-separator theorem for the class of planar graphs. Recently, Bhatt and Leiserson [10] generalized this result by showing that the class of planar graphs has an $\Theta(N^{1/2})$-simultaneous separator. (An N-node graph G is said to have an $f(N)$-*simultaneous separator* if for any 2-coloring (say, black and white) of the nodes of G, there are disjoint subgraphs G_1 and G_2 of G such that G_1 and G_2 each contain $1/2$ of the black nodes and $1/2$ of the white nodes of G, at most $f(N)$ edges link G_1 to G_2, and both G_1 and G_2 have $f(N/2)$-simultaneous separators.) In the following theorem, we show that any N-node graph with an $\Theta(N^{1/2})$-simultaneous separator can be embedded in an

O($NlogN$)-node tree of meshes. As a corollary, we will thus be able to conclude that any N-node planar graph can be embedded in an O($NlogN$)-node tree of meshes.

Theorem 6-1: *Every N-node graph with an $O(N^{1/2})$-simultaneous separator can be embedded in an $O(NlogN)$-node tree of meshes.*

Proof: Let G be an N-node graph with an $f(N)$-simultaneous separator ($f(N)$ will later be chosen to be $O(N^{1/2})$). Partition G into two subgraphs G_1 and G_2 in accordance with the usual separator theorem. Color the nodes of G_1 (G_2) white or black according to whether or not they are linked to a node in G_2 (G_1). (To be precise, we should also weight each node according to the number of nodes in the other subgraph to which it is adjacent.) Now use the simultaneous separator to partition G_1 and G_2. Proceed in this manner until only isolated nodes remain. At each step, color the nodes in the subgraph white if they are adjacent to some node outside of the subgraph and black if they are adjacent only to nodes within the subgraph.

After the first step, at most $f(N)$ edges will link each ($N/2$)-node subgraph to the other. After the second step, at most $f(N)/2 + f(N/2)$ edges will link each ($N/4$)-node subgraph to any other. Using induction, it is not difficult to show that after k steps, at most

$$f(N)/2^{k-1} + f(N/2)/2^{k-2} + f(N/4)/2^{k-3} + \cdots + f(N/2^{k-2})/2 + f(N/2^{k-1})$$

edges will link each ($N/2^k$)-node subgraph to any other. In particular, for $f(N) = O(N^{1/2})$, we can conclude that at most $O(m^{1/2})$ edges will link any m-node subgraph produced by this process to any other subgraph.

Each subgraph produced by the above procedure corresponds in a natural way to a mesh of the tree of meshes. For example, G corresponds to the root mesh, G_1 and G_2 correspond to the second level meshes, and so on. In general, each m-node subgraph corresponds to an $O(m)$-node mesh. Thus each mesh can be used as a switching network to embed the $O(m^{1/2})$ edges which link the corresponding subgraph to other subgraphs. As an example of how this is done, we have included Figure 6-3. In each switching network, the edges entering from the top are linked to the edges entering from the sides. The nodes of G are embedded in the bottom levels of the tree of meshes. $\square$

Corollary 6-1: *Every N-node planar graph can be embedded in an* $O(NlogN)$-*node tree of meshes.*

(b)   Embedding of $M_{2,n}$ in T_{2n}

In fact, any N-node graph with an $O(N^{1/2})$-separator can be embedded in an $O(NlogN)$-node tree of meshes. (See [8] for a proof of this recently discovered fact.) In Section 7.4.3, we prove a slightly weaker result; namely that every N-node graph with an $O(N^{1/2})$-separator can be embedded in *some* $O(NlogN)$-node planar graph.

Of particular importance, however, is the fact that $M_{2,n}$ can be embedded in T_{2n} for any n. For example, consider the embedding of $M_{2,4}$ in T_8 displayed in Figure 6-4. The embedding has been drawn as though it were constructed as part of a larger embedding (say of $M_{2,8}$) in order to illustrate the recursive nature of the general embedding procedure. In addition, the nodes and edges of $M_{2,4}$ have been drawn as they appear in Figure 6-1. For clarity, we have represented the nodes of T_8 as pinpoints and omitted its edges altogether. Also notice that we have not included the bottom two levels of T_8 since they are not needed for the embedding.

The embedding of $M_{2,n}$ in T_{2n} for arbitrary $n \geq 4$ proceeds as follows.

step 1: Remove the roots of the row and column trees of $M_{2,n}$ and all the edges incident to them.

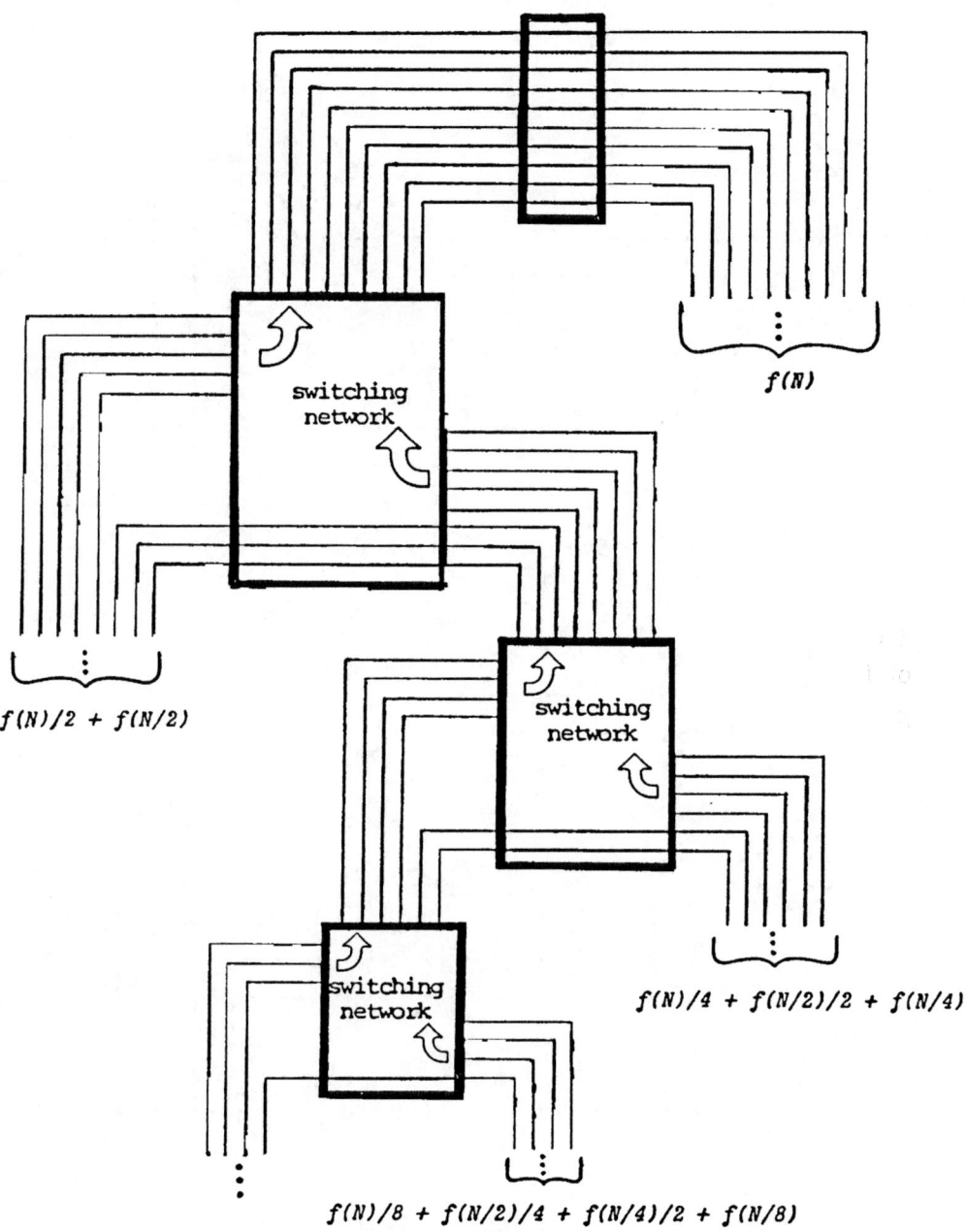

Figure 6-3: *Embedding of a graph with an* $f(N) = O(N^{1/2})$-*simultaneous separator in an* $O(N \log N)$-*node tree of meshes.*

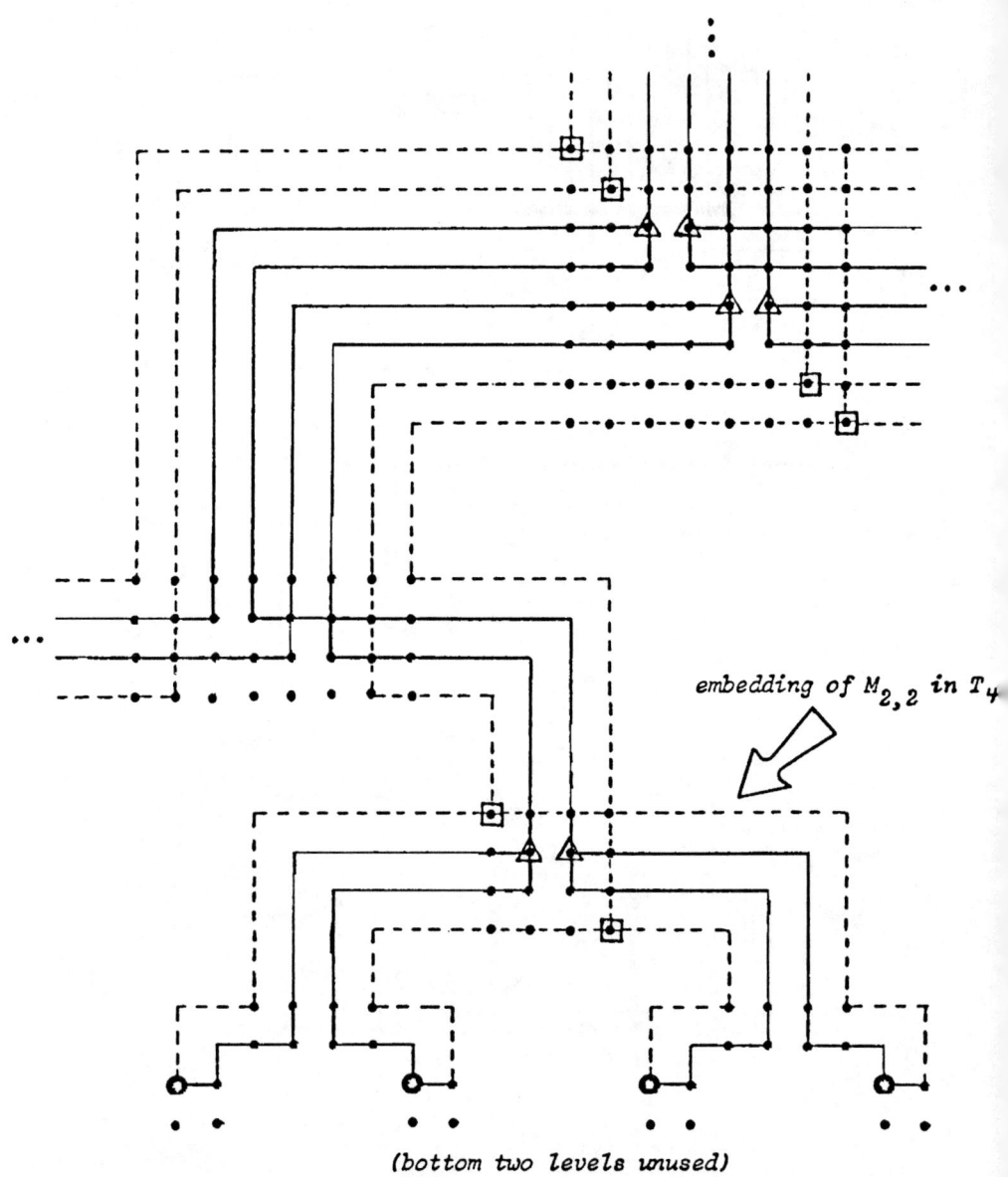

embedding of $M_{2,2}$ in T_4

(bottom two levels unused)

Figure 6-4: *The embedding of $M_{2,4}$ in T_8.*

step 2: Embed the four copies of $M_{2,n/2}$ obtained from step 1 in four separate copies of T_n by calling this procedure recursively.

step 3: Embed the $2n$ roots of the row and column trees in the $2n$ x $2n$ mesh so that

1) the column roots are located at positions (i,i) for $1 \leq i \leq n/2$ and $3n/2 < i \leq 2n$, and

2) the **row roots** are located at positions $(2i-1, 2i-1)$ and $(2i-1, 2i)$ for $n/4 < i \leq 3n/4$

step 4: Draw left and right horizontal edges from each column root to the left and right outer columns of the $2n$ x $2n$ mesh and then to the appropriate node in the top row of the corresponding n x $2n$ mesh. Similarly draw two left edges from each row root with position $(2i-1, 2i-1)$ for some i and two right edges from each row root with position $(2i-1, 2i)$ for some i.

step 5: The n x $2n$ meshes are used as switching networks. In particular, we use them to make the following connections:

1) $(1,i)$ to $(i,1)$ for $1 \leq i \leq n/4$ (column tree connection)

2) $(1,i)$ to $(i+n/2, 1)$ for $n/4 < i \leq n/2$ (column tree connection)

3) $(1, 2i-1)$ to $(i,1)$ for $n/4 < i \leq 3n/4$ (row tree connection)

4) $(1, 2i)$ to $(i, 2n)$ for $n/4 < i \leq 3n/4$ (row tree connection)

5) $(1,i)$ to $(5n/2 - i + 1, 2n)$ for $3n/2 < i \leq 7n/4$ (column tree connection)

6) $(1,i)$ to $(2n - i + 1, 2n)$ for $7n/4 < i \leq 2n$ (column tree connection)

step 6: Each n x $2n$ mesh can be easily linked to two copies of T_n, each of which contains an embedding of $M_{2,n/2}$ produced by this procedure. In particular, attach the wire leaving via the ith row of the n x $2n$ mesh to the node in the ith column of the appropriate *nxn* mesh of T_n for each *n*. (Note that the nodes in the *nxn* meshes are roots of $M_{2,n/2}$ and will become second level nodes of $M_{2,n}$).

6.4   The Augmented Tree of Meshes

As we mentioned in Section 6.3.2, the *N*-node tree of meshes can be laid out so that every wire has length at most O(*logN*). By slightly modifying the graph, however, it is possible to increase the maximum edge length dramatically. The basic idea is to add a complete binary tree with n^2 leaves to the *nxn* tree of meshes so that the leaves of one are linked in a one-to-one fashion with the leaves of the other. It is important that the attachments between the two graphs be made so that the resulting graph (which we call the *nxn augmented tree of meshes* T_n') is planar. For example, we have drawn the *4x4* augmented tree of meshes in Figure 6-5.

It is easily seen that the augmented tree of meshes has, up to a constant, the same bisection width, diameter, separator, layout area and number of nodes as does the original tree of meshes. By adding the binary tree, we have simply decreased the distance between any two *leaves* of the tree of meshes. In Chapter 8, we will show that any layout of the *N*-node tree of meshes has two leaves which are spaced at least $\Omega(N^{1/2}log^{1/2}N)$ apart. We will thus be able to conclude that the maximum edge length of T_n' is at least $\Omega(nlogn) = \Omega(N^{1/2}/log^{1/2}N)$. By applying the techniques developed by Bhatt and Leiserson in [9] to the binary tree portion of the augmented tree of meshes, it is possible to construct a layout that achieves this lower bound for maximum edge length. As the layout itself is not of particular interest, we have not included the details here.

tree of meshes

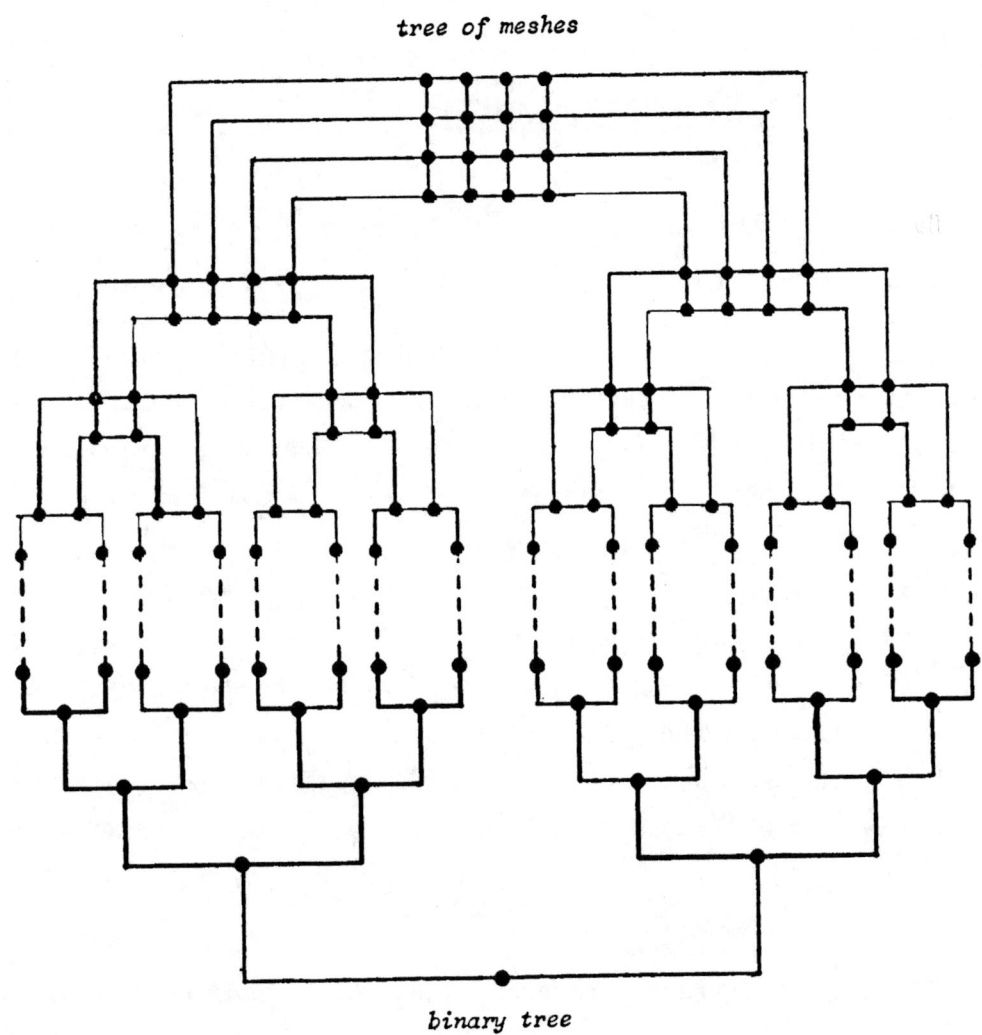

binary tree

Figure 6-5: *The 4x4 augmented tree of meshes $T_4{}'$.*

CHAPTER 7

CROSSING NUMBER ARGUMENTS

In this chapter, we develop a variety of techniques for proving crossing number lower bounds. In the process, we demonstrate the power of crossing number arguments as tools for proving lower bounds on area and maximum edge length. We commence in Section 7.1 by showing that the crossing number of any network is very closely related to its layout area. In Section 7.2, we prove that the crossing number of a network is at least as large (up to a constant factor) as the square of its bisection width. Among other things, this result is used to prove crossing-time tradeoffs for networks which compute transitive functions. In Section 7.3, we describe a more powerful method for proving crossing number lower bounds. The latter method is the only method known for proving nontrivial lower bounds for meshes of trees. We conclude in Section 7.4 with a collection of miscellaneous results.

7.1 The Relationship Between Crossing Number and Layout Area

It is clear that the crossing number of a graph is a lower bound on its layout area. In what follows, we show that the crossing number is also close to an upper bound for layout area. This simple result has been observed by a number of people, including Lipton and Tarjan [60] and Leiserson [55]. Similar techniques have been used by Jia-Wei and Rosenberg in [35].

Theorem 7-1: *Given an optimal drawing D for an N-node graph G with crossing number C, it is possible to construct a layout for G with area at most* $O((C+N)log^2(C+N))$.

Proof: Let D be a drawing of G in the plane with C crossings. Replace each crossing of D with an *artificial node*. The resulting graph G' has $C+N$ nodes and is planar. Using the methods developed by Leiserson [55] and Valiant [98], G' can be laid out in $O((C+N)log^2(C+N))$ area. Once this is done, it is a simple matter to replace the artificial nodes with their original wire crossings to obtain the desired layout for G. □

The previous theorem also demonstrates the importance of finding a good layout strategy for planar graphs. The relationship between laying out general graphs and laying out planar graphs is formalized in the following corollary to Theorem 7-1.

Corollary 7-1: *If every N-node planar graph can be laid out in A(N) area, then every N-node graph with crossing number C can be laid out in* $O(A(C+N))$ *area.*

7.2 Lower Bounds Based on Bisection Width

As was mentioned in Chapter 5, the standard techniques for proving area lower bounds proceed by first proving bisection width lower bounds and then applying the inequality $A \geq \Omega(B^2)$ from Theorem 5-1. In what follows, we prove a similar inequality for crossing numbers. As a result, we are able to extend a number of area lower bounds to crossing lower bounds. Of particular interest is the crossing-time tradeoff proved in Section 7.2.2.

7.2.1   Crossing Number Lower Bounds

In what follows, we prove the analogous result of Theorem 5-1 for crossing numbers.

Theorem 7-2: *If G is an N-node graph with crossing number C and bisection width B, then* $C + N \geq \Omega(B^2)$.

Proof: Let D be a drawing of G in the plane with C crossings. As in Theorem 7-1, replace each crossing of D with an artificial node. Call the resulting planar graph G' and note that it has precisely $C + N$ nodes. Using the weighted version of the Lipton-Tarjan planar separator theorem [60], it is possible to bisect the real nodes of G' (by assigning weight 1 to the real nodes and weight 0 to the artificial nodes) without cutting more than $O((C+N)^{1/2})$ edges. After replacing the artificial nodes with their original edge crossings, it becomes apparent that we have, in fact, constructed an $O((C+N)^{1/2})$-bisection for G. By squaring, we find that $C + N \geq \Omega(B^2)$. □

Theorem 7-2 has a number of immediate and important consequences. For example, virtually every nonlinear area lower bound can now be transformed into a crossing number lower bound. In particular, Theorem 7-2 provides a good lower bound for the crossing number of the shuffle-exchange graph. This bound is stated in the following corollary.

Corollary 7-2: *Every layout for the N-node shuffle-exchange graph has at least* $\Omega(N^2/\log^2 N)$ *wire crossings.*

Proof: Thompson [93] showed that the N-node shuffle-exchange graph has bisection width $\Omega(N/\log N)$. By Theorem 7-2, this means that the shuffle-exchange graph has C crossings where $C + N \geq \Omega(N^2/\log^2 N)$. Since $N = o(N^2/\log^2 N)$, this means that $C \geq \Omega(N^2/\log^2 N)$, as claimed. □

7.2.2   Crossing-Time Tradeoffs

In [93], Thompson showed that the area A and time T required to compute any N-variable transitive function (such as fast Fourier transform and sorting) must satisfy $AT^2 \geq \Omega(N^2)$. This fundamental result is based on bisection width arguments. As a result, we can generalize the tradeoff to crossing numbers by applying Theorem 7-2. The result is stated in the following theorem.

Theorem 7-3: *Any chip which computes an N-variable transitive function in time $T=o(N^{1/2})$ must have at least C wire crossings where $CT^2 \geq \Omega(N^2)$.*

Proof: Using straightforward information transfer arguments, Thompson [93] showed that any network which computes an N-variable transitive function in time T must have bisection width $\Omega(N/T)$. If $T = o(N^{1/2})$, then $N = o(N^2/T^2)$ and we can apply Theorem 7-2 to show that the network has $C \geq \Omega(N^2/T^2)$ crossings. □

Note that we could reprove Corollary 7-2 as a corollary to Theorem 7-3 since the N-node shuffle-exchange graph can compute a variety of N-variable transitive functions in $O(logN)$ time. More generally, Theorem 7-3 states that any network which computes transitive functions quickly must have many wire crossings. In addition, Theorem 7-3 can be extended in much the same way that Thompson's original area-time tradeoff result has been extended by numerous others [1, 5, 20, 21, 37, 38, 59, 74, 75, 85, 99, 100, 101].

7.2.3   Edge Crossing Lower Bounds

Given a lower bound for the crossing number of a graph, it is not difficult to prove a lower bound for the maximum edge crossing of the graph. For example, if an N-node graph has C crossings, then some edge must cross at least $\Omega(C/N)$ other edges. This fact can be used to prove that any layout for the N-node shuffle-exchange graph contains a wire which crosses at least $\Omega(N/log^2 N)$ other wires. In general, we have the following corollary to Theorem 7-2 for maximum edge crossing. Note that this bound also holds for maximum edge length.

Corollary 7-3: *Any N-node graph with bisection width* $B \gg \Omega(N^{1/2})$ *has maximum edge crossing at least* $\Omega(B^2/N)$.

7.3   More Sophisticated Lower Bounds

In this section, we describe a general method for proving crossing number lower bounds. For some graphs (such as the shuffle-exchange graph), the method gives the same lower bounds as Theorem 7-2. For other graphs (such as the mesh of trees), the method is substantially more powerful than Theorem 7-2. A variant of the method will be used in Chapter 8 to prove lower bounds for wire area.

7.3.1   Description of the Method

Given a drawing D for an N-node connected graph G, we will construct a drawing D' for the complete graph on N nodes K_N by tracing over the edges of D. For example, we have done this for the 4-node graph shown in Figure 7-1. The edges of the original graph are drawn with dashed lines while solid lines indicate edges of K_4 .

If we are careful not to trace over each edge of D too many times during the construction of D', it may be possible to infer something about the number of crossings in D by counting the number of crossings in D'. This is due to the fact that the number of crossings in D is closely related to the number of crossings in D'. For example, if e_1 and e_2 are edges of G which cross in D and e_1 is traced over s_1 times while e_2 is traced over s_2 times, then the crossing of e_1 with e_2 will appear $s_1 s_2$ times in D'. Such a crossing of D' is called a *crossing of the first kind*.

For example, there are four crossings of the first kind in the drawing of K_4 in
Figure 7-1.

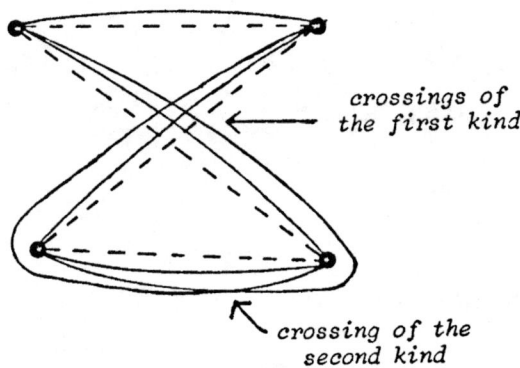

$$crossings\ of$$
$$the\ first\ kind$$

$$crossing\ of\ the$$
$$second\ kind$$

Figure 7-1: *Construction of K_4 from the drawing of a 4-node graph.*

Sometimes, it is necessary for two edges of D' to cross while traversing the
same edge of D. Such a crossing is called a *crossing of the second kind*. Note that
there is only one crossing of the second kind in the drawing of K_4 in Figure 7-1.
Since D' can easily be drawn so that no pair of edges cross each other more than
once, there are usually not very many crossings of the second kind. More
precisely, if G has edges $e_1, \ldots, e_k$ and if edge e_i is traced over s_i times for each
i during the construction of D', then D' can have at most $\sum_{i=1}^{k} s_i^2/2$ crossings of
the second kind. For most applications of the method, this number is substantially
smaller than the number of crossings of the first kind in D' and thus we usually do
not have to worry about crossings of the second kind.

By showing that the number of crossings in D' is large, we can conclude that
there must be a large number of crossings in D. For example, if each edge of D is
traced over at most s times during the construction of D' and D' is found to have

C crossings, then we can conclude that D has at least C/s^2 crossings. This follows from the fact that each crossing of D is replicated at most s^2 times in D'. (Note that we have neglected crossings of the second kind in this argument.)

Fortunately, it is easy to find a good lower bound on the number of crossings in any drawing of K_N. We state the result formally in the following lemma. The proof can also be found in Kleitman's work [39] but is generally regarded as folklore.

Lemma 7-1 (Kleitman [39]): *The crossing number of K_N, the complete graph on N nodes, is at least $N(N-1)(N-2)(N-3)/120$ for $N \geq 5$.*

Proof: Let D be a drawing of K_N in the plane with the smallest possible number of crossings $C(N)$. We may assume that no pair of edges which cross in D are incident to a common node. Otherwise, it would be possible to produce a drawing D' for K_N with $C(N)-1$ crossings by exchanging the parts of the crossing edges which lie between the common node and the point of crossing. This would contradict the minimality of $C(N)$.

Consider the N subdrawings of D obtained by deleting one of the nodes and all of the edges incident to it. Note that each crossing of D appears in precisely $N-4$ of the subdrawings. (A crossing does not appear in any of the 4 subdrawings which correspond to the deletion of a node incident to an edge of the crossing.) Since each of the subdrawings is a drawing of K_{N-1}, each must have at least $C(N-1)$ crossings. Thus $(N-4)C(N) \geq NC(N-1)$. Applying the inequality recursively and noting that $C(5)=1$, we can conclude that

$$C(N) \geq [N/(N-4)][(N-1)/(N-5)] \cdots [6/2]$$
$$= N(N-1)(N-2)(N-3)/120 \quad \text{for } N \geq 5 . \ \square$$

7.3.2   Lower Bounds for the Shuffle-Exchange Graph

In what follows, we show how to use the method described in Section 7.3.1 to prove a crossing number lower bound for the shuffle-exchange graph. Although the same bound was proved as Corollary 7-2, the following proof is purely combinatorial and does not rely on information transfer arguments. In addition, the proof nicely illustrates how the method works in general.

Let D be any drawing of the N-node shuffle-exchange graph G where $N = 2^k$. We first show how to construct a drawing D' of K_N on the nodes of G without tracing over any edge of D more than $N\log N$ times.

Given any pair of nodes $a_k \cdots a_1$ and $b_k \cdots b_1$, draw the edge from $a_k \cdots a_1$ to $b_k \cdots b_1$ along the path

$$a_k \cdots a_3 a_2 a_1 \longrightarrow a_k \cdots a_3 a_2 b_1 \longrightarrow b_1 a_k \cdots a_3 a_2 \longrightarrow b_1 a_k \cdots a_3 b_2 \longrightarrow$$
$$b_2 b_1 a_k \cdots a_3 \longrightarrow \cdots \longrightarrow b_{k-1} \cdots b_2 b_1 b_k \longrightarrow b_k b_{k-1} \cdots b_2 b_1.$$

(In order that every edge of K_N not be drawn twice, we should assume that the value of $a_k \cdots a_1$ is less than that of $b_k \cdots b_1$ but this has no bearing on the argument.)

Wherever $a_i = b_i$ for some i, the preceding path will have a loop. When actually drawing the edges of D', we ignore such loops. For example, the edge from 01100 to 11101 is drawn along the path

$$01100 \xrightarrow{e} 01101 \xrightarrow{s} 10110 \xrightarrow{s} 01011 \xrightarrow{s} 10101 \xrightarrow{s} 11010 \xrightarrow{e}$$
$$11011 \xrightarrow{s} 11101.$$

For convenience, we have labeled the shuffle edges with an $\xrightarrow{s}$ and the exchange edges with an $\xrightarrow{e}$. Note also that we have omitted loops at 10110, 01011 and 10101.

It is not difficult to show that every edge of D is traced over at most $N\log N$

times during the construction of D'. For example, consider the shuffle edge linking $a_k \cdots a_2 a_1$ to $a_1 a_k \cdots a_2$. It is traced over during the construction of edges of D' which link a node of the form

$$a_{k-i+1} \cdots a_2 \overbrace{* \cdots *}^{k-i}$$

to a node of the form

$$\overbrace{* \cdots *}^{i} a_1 a_k \cdots a_{k-i+2}$$

for any i, $1 \leq i \leq k$ (where $*$ indicates either a 0-bit or a 1-bit). It is easily seen that there are at most $k2^k$ such edges in D' and thus each shuffle edge is traced over at most $N log N$ times. A similar argument shows that each exchange edge is also traced over at most $N log N$ times.

Since each edge is traced over at most $N log N$ times, there can be at most

$$(3N/2)\,[(N log N)^2/2]\ =\ 3N^3/(4 log^2 N)$$

crossings of the second kind in D'. This is substantially less than total number $\Omega(N^4)$ of crossings in D'. Thus D' must have $\Omega(N^4)$ crossings of the first kind. As each edge of D is traced over at most $N log N$ times, this means that D has at least $\Omega(N^4/(N log N)^2) = \Omega(N^2/log^2 N)$ crossings.

It is worth pointing out that the preceding argument can also be used to prove that the N-node shuffle-exchange graph has bisection width at least $\Omega(N/log N)$. The result follows from the observation that K_N has bisection width $O(N^2)$ and the fact that every edge of D was traced over at most $N log N$ times during the construction of D'. This means that the bisection width of the N-node shuffle-exchange graph is at least $\Omega(N^2/(N log N)) = \Omega(N/log N)$, as claimed.

In fact, a similar modification of the method described in Section 7.3.1 can be used to find tight bisection width lower bounds for *all* of the networks we have investigated. For most of these networks, however, it is much more useful to study the corresponding crossing number and wire area bounds.

7.3.3 Lower Bounds for the 2-Dimensional Mesh of Trees

In this section, we use a more sophisticated version of the method described in Section 7.3.1 to prove a nontrivial lower bound on the crossing number of the 2-dimensional mesh of trees.

Theorem 7-4: *The crossing number of the N-node 2-dimensional mesh of trees is at least* $\Omega(NlogN)$.

Proof: As before, let $M_{2,n}$ denote the 2-dimensional mesh of trees (where n is a power of 2). We will show that the crossing number of $M_{2,n}$ is at least

$$(n^2logn - 121n^2 + 121n)/40 \quad \text{for all} \quad n \geq 1.$$

Since $M_{2,n}$ has $N = \Theta(n^2)$ nodes, this will be sufficient to prove the desired result.

The proof consists of two steps. In the first, we show how to construct a drawing of K_{n^2} from any drawing of $M_{2,n}$ by tracing over the edges of $M_{2,n}$. We then apply Lemma 7-1 to conclude that there are a large number of crossings among the edges in the top levels of the binary trees of $M_{2,n}$. In the second step, we complete the proof by inductively applying the result of the first step.

step 1: Let D be any drawing of $M_{2,n}$ in the plane. From this drawing, we can construct a drawing D' of K_{n^2} in the following way. First locate the n^2 leaves of the binary trees of D. They will serve as the nodes for K_{n^2}. Given any pair (i,j) and (k,l) of these nodes, draw an edge from (i,j) to (k,l) along the unique path from (i,j) to (i,l) in the *ith* row tree of D and then from (i,l) to (k,l) in the *lth* column tree of D. (In order that each edge not be drawn twice, we shall assume that $i \leq k$ and, when $i = k$, that $j < l$.) As usual, we assume that the edges of D' are drawn so that no pair cross each other more than once.

We next count the number of crossings of the second kind in D'. In order to do this, we need to count the number of times each edge of D is traced over during the construction of D'. It is not difficult to show that each edge in the ith level of a binary tree of $M_{2,n}$ (henceforth, referred to as a *type i* edge) is traced over at most

$$n2^{-i}(n^2 - n^2 2^{-i}) \leq n^3 2^{-i}$$

times for any $i \leq logn$ during the construction of D'. Thus at most $n^6 2^{-2i-1}$ crosses of the second kind can occur at any type i edge of D. Since there are $2^{i+1}n$ type i edges in $M_{2,n}$, we can conclude that the total number of crosses of the second kind in D' is at most

$$\sum_{i=1}^{logn}(2^{i+1}n)(n^6 2^{-2i-1}) = n^7 \sum_{i=1}^{logn} 2^{-i} \leq n^7$$

We next count the number of crossings of the first kind (i.e., those corresponding to crosses in D). We say that a crossing of D is *type i-j* if it is the crossing of a type i edge and a type j edge. Let t_{ij} denote the number of type i-j crossings in D and set $t_i = \sum_{j=i}^{logn} t_{ij}$. Since each type i edge is traced over at most $n^3 2^i$ times, each type i-j crossing of D produces at most $(n^3 2^{-i})(n^3 2^{-j}) = n^6 2^{-i-j}$ crosses of the first kind in D'. Thus the total number of crossings of the first kind in D' is at most

$$\sum_{i=1}^{logn} \sum_{j=i}^{logn} n^6 2^{-i-j} t_{ij} \leq n^6 \sum_{i=1}^{logn} 2^{-2i} t_i.$$

Summing, we find that the total number of crossings of either kind in D' is at most $n^7 + n^6 \sum_{i=1}^{logn} 2^{-2i} t_i$ By Lemma 7-1, this number must be at least $n^2(n^2-1)(n^2-2)(n^2-3)/120$ for $n^2 \geq 5$. Simplifying, we can conclude that

$$\sum_{i=1}^{logn} 2^{-2i} t_i \geq (n^2 - 121n)/120 \quad \text{for } n \geq 6.$$

Let $s_k = \sum_{i=1}^{k} t_i$ be the **number of crossings involving at least one edge from the top k levels of some** binary tree of $M_{2,n}$. In what follows, we will use the preceding inequality to show that $s_k \geq (n^2 - 121n)k/40$ for at least some value of $k \geq 1$. Assume otherwise and observe that

$$\sum_{i=1}^{logn} 2^{-2i}t_i = \sum_{i=1}^{logn} 2^{-2i}(s_i\text{-}s_{i\text{-}1})$$

where s_0 is defined to be 0. The coefficient of each s_i $(i > 0)$ in this sum is $2^{-2i}\text{-}2^{-2i\text{-}2}$ which is positive so for each i we may substitute $(n^2\text{-}121n)i/40$ as an upper bound for s_i in order to see that

$$\sum_{i=1}^{logn} 2^{-2i}t_i < [(n^2\text{-}121n)/40] \sum_{i=1}^{logn} 2^{-2i}[i\text{-}(i\text{-}1)]$$

$$= [(n^2\text{-}121n)/40] \sum_{i=1}^{logn} 4^{-i}$$

Since $\sum_{i=1}^{logn} 4^{-i} \leq 1/3$ for all n, we can conclude that

$$\sum_{i=1}^{logn} 2^{-2i}t_i < (n^2\text{-}121n)/120 \quad \text{for all } n>121,$$

a contradiction. Thus for all $n\geq 121$, there is a $k\geq 1$ such that $s_k \geq (n^2\text{-}121n)k/40$.

step 2: Let $C(n)$ denote the crossing number of $M_{2,n}$. Using the result of step 1, we will now show by induction on n that $C(n) \geq (n^2 logn - 121n^2 + 121n)/40$ for all $n\geq 1$.

As $(n^2 logn - 121n^2 + 121n)/40$ is nonpositive for small n, the lower bound trivially holds for all $n<128$. Assume that the lower bound holds for all $m<n$ where $n\geq 128$ and let D be any drawing for $M_{2,n}$. By counting the crossings of D in two groups according to whether or not at least one edge of the crossing is contained in the top k levels of the binary trees of $M_{2,n}$, we can observe that

$$C(n) \geq 2^{2k}C(n2^{-k}) + s_k.$$

(Recall the definition of s_k and the structure of $M_{2,n}$.) By choosing k as in step 1 so that $s_k \geq (n^2\text{-}121n)k/40$ and applying the inductive hypothesis for $C(n2^{-k})$, we obtain

$$C(n) \geq 2^{2k}[n^2 2^{-2k}(logn-k)/40 - 121n^2 2^{-2k}/40 + 121n2^{-k}/40] + n^2 k/40 - 121nk/40$$

$$\geq n^2 logn/40 - 121n^2/40 + 121n/40 + 121n(2^k\text{-}k\text{-}1)/40$$

$$\geq (n^2 logn - 121n^2 + 121n)/40 .$$

Thus the inductive hypothesis is established and we can conclude that the crossing number of $M_{2,n}$ is at least $\Omega(n^2 \log n) = \Omega(N \log N)$. $\square$

In Section 7.4.3, we will show that the crossing number of any N-node graph with an $O(N^{1/2})$-separator is at most $O(N \log N)$. Thus, we will be able to conclude that the crossing number of the N-node 2-dimensional mesh of trees is precisely $\Theta(N \log N)$.

7.3.4   Lower Bounds for the r - Dimensional Mesh of Trees

By modifying the proof of Theorem 7-4, it can be shown that any layout of the r - dimensional mesh of trees must have very long wires. In particular, they must be as long as the width of any optimal layout for the graph. We state this result more precisely in the following theorem.

Theorem   7-5:   *Any drawing of the N-node r - dimensional mesh of trees contains an edge which crosses at least $\Omega(N^{1-1/r})$ other edges.*

Proof: The r - dimensional $\overbrace{n x n x \cdots x n}^{r}$ mesh of trees $M_{r,n}$ has $N = (r+1)n^r - rn^{r-1} = \Theta(n^r)$ nodes for bounded r. We will show that any layout D of $M_{r,n}$ contains an edge which crosses at least $\Omega(n^{r-1}) = \Omega(N^{1-1/r})$ other edges, thus proving the theorem. The method used is very similar to that of Theorem 7-4.

As we did for the case of $r=2$ in Theorem 7-4, we first construct a drawing D' of the complete graph on the n^r leaves of $M_{r,n}$. Each type i edge of D is traced over at most $n^{r+1} 2^{-i}$ times by this procedure. Thus the total number of crossings in D' is at most

$$(rn^{3r+1})/2 \; + \; n^{2r+2} \sum_{i=1}^{\log n} 2^{2i} t_i$$

where, as before, $t_i = \sum_{j=i}^{\log n} t_{ij}$ and t_{ij} is the number of type i-j crossings in D. Applying Lemma 7-1, we can conclude that $\sum_{i=1}^{\log n} 2^{2i} t_i \geq \Omega(n^{2r-2})$.

Let $s_k = \sum_{i=1}^{k} t_i$ be the total number of crossings of D involving an edge from the top k levels of the binary trees in $M_{r,n}$. Using arguments similar to those used to prove Theorem 7-4, it is not difficult to show that for large n, there exists a k such that $s_k \geq \Omega(n^{2r-2} 2^k)$. As there are only $rn^{r-1}(2^{k+1}-2)$ edges in the top k levels of $M_{r,n}$ for any k, we can conclude that at least one of them crosses at least $\Omega(n^{r-1})$ other edges. $\square$

It is worth pointing out that the preceding arguments can also be used to show that the crossing number of the N-node r - dimensional mesh of trees is $\Theta(N^{2-2/r})$ for bounded $r > 2$.

7.4   Further Methods

In this section, we describe some additional methods for proving crossing number bounds. We first generalize Lemma 7-1 to prove a combinatorial lower bound on the crossing number of any N-node graph with at least $4N$ edges. This result is then used in Section 7.4.2 to prove crossing number lower bounds for a class of graphs which are similar to the 2-dimensional mesh of trees. We conclude by proving a nontrivial upper bound on the crossing number of graphs which have $O(N^{1/2})$-separators. As a corollary, we show that any N-node graph with an $O(N^{1/2})$-separator can be embedded in some $O(N\log N)$-node planar graph, thus generalizing Theorem 6-1.

7.4.1   A Combinatorial Lower Bound for Crossing Numbers

In this section, we substantially generalize the result of Lemma 7-1. Throughout, we assume that G is a *simple* graph (i.e., that it has no loops or multiple edges).

Theorem 7-6: *If G is a graph with E edges and N nodes where $E \geq 4N$, then the crossing number of G is at least $E^3/375N^2$.*

Proof: The proof is by induction on N. For $N=1$, the result is vacuously true. Assume that the result is true for all $N' < N$ where $N > 1$ and let G be a graph with N nodes and E edges where $E \geq 4N$. We will show that the crossing number c of G is at least $E^3/375N^2$, thus proving the theorem. There are two cases to consider.

case 1: $4N \leq E < 5N$

We first use Euler's formula [11] in order to show that the genus of G is large. Euler's formula states that

$$E + 2 = N + f + 2g$$

where f is the number of faces of any proper embedding of G on a surface of genus g. Since G has no loops or multiple edges, every face contains at least 3 edges and thus $3f \leq 2E$. Substituting, we find that

$$2g = E + 2 - N - f$$
$$\geq E + 2 - N - (2E/3)$$
$$= E/3 + 2 - N$$

and thus that $g > (E - 3N)/6$. For $4N \leq E < 5N$, it is not difficult to show that $(E - 3N)/6 \geq E^3/375N^2$ and thus that $g \geq E^3/375N^2$.

Given any graph with crossing number C, it is possible to find a proper embedding of the graph on a surface with genus C. We can do this by drawing the graph on a sphere so that only C pairs of edges cross and then putting a "handle" in the region immediately surrounding each crossing. The edges of the crossing can then be redrawn through the handle so that they no longer cross. As the resulting surface has genus C, we can conclude that $g \leq C$ for any graph with genus

g and crossing number C. This means that $C \geq E^3/375N^2$ for G.

case 2: $E \geq 5N$

Let $d_1, \ldots, d_N$ be the degrees of the N nodes of G and let D be an optimal drawing of G. As usual, we can assume that no pair of edges which cross in D are incident to the same node of G. Consider the subdrawing D_i of D obtained by deleting the ith node of G and all the edges incident to it. This subdrawing is also a drawing of a graph with N-1 nodes and E-d_i edges. Since $E \geq 5N$ and $d_i \leq N$-1, we can conclude that

$$E - d_i \geq 4N + 1 > 4(N\text{-}1).$$

Thus we can apply the inductive hypothesis to D_i in order to conclude that it has at least $(E\text{-}d_i)^3/[375(N\text{-}1)^2]$ crossings.

Each crossing of D will appear in precisely N-4 of the N subdrawings of D produced by the above procedure. Applying the technique used to prove Lemma 7-1, we can thus conclude that

$$C \geq [1/(N\text{-}4)] \sum_{i=1}^{N} (E\text{-}d_i)^3/[375(N\text{-}1)^2]$$

$$= [1/375(N\text{-}4)(N\text{-}1)^2] \sum_{i=1}^{N} (E^3 - 3E^2 d_i + 3E d_i^2 - d_i^3)$$

$$= [1/375(N\text{-}4)(N\text{-}1)^2] [E^3 N - 3E^2(2E) + \sum_{i=1}^{N} (3E d_i^2 - d_i^3)]$$

Since $2E = \sum_{i=1}^{N} d_i$, it is not difficult to show that $\sum_{i=1}^{N} (3E d_i^2 - d_i^3)$ attains its minimal value when $d_i = 2E/N$ for $1 \leq i \leq N$. At this point,

$$\sum_{i=1}^{N} (3E d_i^2 - d_i^3) \geq 12E^3/N - 8E^3/N^2$$

and thus

$$C \geq (E^3 N - 6E^3 + 12E^3/N - 8E^3/N^2) / [375(N^3 - 6N^2 + 9N - 4)].$$

For $N \geq 2$, this expression can easily be reduced to show that $C \geq E^3/375N^2$. $\square$

It is interesting to note that the lower bound proved in Theorem 7-6 is (up to a constant) tight. For example, the N-node graph consisting of N^2/E disjoint copies of $K_{E/N}$ has $O(E)$ edges and crossing number at most $O(E^3/N^2)$ for any $E \geq 4N$.

7.4.2   Applications

When defining the 2-dimensional mesh of trees, we required that the binary trees be interconnected so that $M_{2,n}$ contain 2^{2k} disjoint copies of $M_{2,n2^k}$ as subgraphs for any k. Not only is this definition the most natural, but it also allows us to use induction in the lower bound proofs for the network. Surprisingly, however, the constraint is not necessary in order to show that $M_{2,n}$ can perform matrix-vector multiplication, sorting or switching in $O(\log n)$ time. In fact, any network consisting of n row trees and n column trees which share the same set of leaves can do these operations quickly. Thus it is conceivable that some other arrangement of the tree interconnections might lead to a network with a smaller crossing number. In what follows, we use Theorem 7-6 to show that this is not the case.

Theorem 7-7: *If G is an N-node graph formed in the same way as the $n \times n$ mesh of trees except that arbitrary interconnections are allowed between the leaves of the binary trees, then G must have crossing number at least $\Omega(N \log N)$.*

Proof: Let G_k denote the subgraph of G obtained by deleting the nodes and edges in the top k levels of the binary trees of G for $0 \leq k < \log n$. For example, if $G \simeq M_{2,n}$, then G_k consists of 2^{2k} disjointcopies of $M_{2,n2^k}$. Otherwise, G_k is a graph for which each node of the original $n \times n$ matrix of nodes is a leaf of a horizontal complete binary tree of depth $\log n - k$ and a leaf of a vertical complete binary tree of depth $\log n - k$. For each k, let H_k denote the graph whose nodes are the n^2 leaves of G_k and whose edges are the paths in G_k of the form

leaf − path in horizontal binary tree − leaf − path in vertical binary tree − leaf.

Note that if $G \simeq M_{2,n}$, then H_k consists of 2^{2k} disjoint copies of $K_{n^2 2^{-2k}}$. In any case, H_k is a regular graph for which each node has degree $n^2 2^{-2k-1}$

Given any drawing D_k of G_k , it is easy to construct a drawing $D_k{}'$ for H_k by tracing over the edges of G_k in the natural way. It is not difficult to see that each type i edge of G is traced over at most $(2^{\log n - k})^3 2^{(i-k)} = n^3 2^{-2k-i}$ times by this procedure for $i > k$. Thus each type i-j crossing is reproduced at most $n^6 2^{-4k-i-j} \leq n^6 2^{-4k-2i}$ times for $j \geq i > k$.

Given any drawing D of G, construct 2^{6k} separate drawings $D_k{}'$ of H_k for each $k \geq 0$.   Each type i-j crossing of D will appear a total of

$$\sum_{k=0}^{H}(n^6 2^{-4k-2i})(2^{6k}) = n^6 2^{-2i}\sum_{k=0}^{H}2^{2k}$$
$$\leq O(n^6)$$

times in these drawings. In what follows, we will show that there are at least $\Omega(n^8 \log n)$ total crossings of the first kind in these drawings. We will thus be able to conclude that the crossing number of G is at least $\Omega(n^2 \log n)$.

As H_k has $E_k = O(n^4 2^{-2k})$ edges and $N_k = n^2$ nodes, we can apply Theorem 7-6 to conclude that $D_k{}'$ has at least $\Omega(E_k{}^3/N_k{}^2) = \Omega(n^8 2^{-6k})$ crossings. Thus there are at least $\Omega(n^8)$ crossings among the 2^{6k} drawings $D_k{}'$. Summing over k for $0 \leq k \leq \log n$, we find that there are at least $\Omega(n^8 \log n)$ total crossings among all of the drawings $\{D_k{}' \mid 0 \leq k \leq \log n \}$. It is not difficult to check that there are at most $O(n^7 2^{-5k})$ crossings of the second kind in each drawing of H_k . As there are 2^{6k} such drawings for each k, we can conclude that there are at most

$$\sum_{k=1}^{\log n}(n^7 2^{-5k})2^{6k} \leq O(n^8)$$

total crossings of the second kind in all the drawings $\{D_k{}' \mid 0 \leq k \leq \log n \}$. Thus there are at least $\Omega(n^8 \log n)$ total crossings of the first kind and the crossing number of G is at least $\Omega((n^8 \log n)/n^6) = \Omega(n^2 \log n) = \Omega(N \log N)$. $\square$

As a corollary, we can see once again that the crossing number of $M_{2,n}$ is at least $\Omega(N log N)$.

7.4.3   An Upper Bound for Crossing Numbers

Since any N-node graph with an $O(N^\alpha)$-separator for some $\alpha > 1/2$ has an $O(N^{2\alpha})$-area layout, we can easily see that it also has crossing number at most $O(N^{2\alpha})$. By Theorem 7-2, we can conclude that this bound is tight since many such graphs also have bisection width at least $\Omega(N^\alpha)$.

The situation is not as clear for graphs with $O(N^{1/2})$-separators, however. For example, the best known upper bound on the layout area of an N-node graph with an $O(N^{1/2})$-separator is $O(N log^2 N)$ yet no such graph is known to have a crossing number greater than $\Omega(N log N)$. In what follows, we prove a tight upper bound on the crossing number of any such graph.

Theorem   7·8:   *The crossing number of any N-node graph with an $O(N^{1/2})$-separator is at most* $O(N log N)$.

Proof: Given such a graph G, we will construct a drawing for G with at most **$O(N log N)$** crossings.   In order to construct the drawing, we will

1) decompose G into subgraphs according to the separator theorem,

2) draw the subgraphs by recursively calling the procedure, and

3) draw the edges which link the subgraphs together without introducing too many crossings and so that every node remains "close" to the exterior of the drawing.

In order to illustrate the procedure, we will describe in detail how drawings D_1 and D_2 of two m-node subgraphs are used to construct a drawing D of the combined $2m$-node subgraph. Let $C(m)$ denote number of crossings in D_1 or D_2, whichever is larger. Further let $d(m)$ denote the maximum number of edges which must be crossed in order to draw an edge from any node in D_1 or D_2 to the exterior of D_1 and D_2. Construct D from the drawings of D_1 and D_2 by drawing in the $O(m^{1/2})$ edges which link them together in the best way possible. Now let $C(2m)$ and $d(2m)$ be the obvious values for the constructed drawing D. It is not difficult to show that

$$C(2m) \leq 2C(m) + O(m) + O(m^{1/2}d(m))$$

and that

$$d(2m) \leq d(m) + O(m^{1/2}).$$

Solving the recurrences in general, we find that $d(m) \leq O(m^{1/2})$ and thus that $C(m) \leq O(m \log m)$. Thus the above procedure can be used to find a drawing for G with at most $O(N \log N)$ crossings. $\square$

Using the preceding result, we can substantially generalize Theorem 6-1.

Theorem 7-9: *Any N-node graph with an $O(N^{1/2})$-separator can be embedded in an $O(N \log N)$-node planar graph.*

Proof: Construct a drawing of the graph with $O(N \log N)$ crossings according to the method described in the proof of Theorem 7-8. Replace each edge crossing in the drawing with an artificial node. The resulting graph has $O(N \log N)$ nodes, is planar and embeds the original graph. $\square$

CHAPTER 8

WIRE AREA ARGUMENTS

In this chapter, we extend the method of Section 7.3 to prove lower bounds on the wire area of a variety of networks. In each proof, we use a layout of a network to produce a layout for the complete graph. By showing that the nodes of the layout are widely spread out, we will be able to conclude that the wire area of the layout for the complete graph is very large. Provided that the edges of the original network were not traced over too many times, we can then reason that the wire area of the original network is also large.

8.1 Lower Bounds for the 2-Dimensional Mesh of Trees

In this section, we find tight lower bounds for the layout area and maximum edge length of the 2-dimensional mesh of trees.

Theorem 8-1: *The wire area of the N-node 2-dimensional mesh of trees is at least* $\Omega(N log^2 N)$.

Proof: As usual, we denote the $n \times n$ mesh of trees by $M_{2,n}$. In addition, let $W(n)$ denote the wire area of $M_{2,n}$ and let α be a positive constant such that

$(*)$ $\alpha \leq n/(4 log^2 n)$ for all $n \geq 2$, and

$(**)$ $\alpha \leq 2^{2i-20}/(\beta^2 i^6)$ for all $i \geq 1$

where $\beta = \sum_{j=1}^{\infty} j^{-2}$, also a constant. Clearly such a constant exists ($\alpha = 2^{-30}$ should suffice) and clearly $W(n) \geq \alpha n^2 log^2 n$ for $n=1$ and 2. Consider a value of $n \geq 4$ which is a power of 2 and assume that for all values of $m < n$ which are powers 2 that $W(m) \geq \alpha m^2 log^2 m$. We will use induction to show that $W(n) \geq \alpha n^2 log^2 n$. Since $M_{2,n}$ has $N = O(n^2)$ nodes, this will be sufficient to prove the theorem.

Consider any layout for $M_{2,n}$ which uses $W(n)$ wire. Partition the layout into three vertical strips V_0, V_1 and V_2 so that the center strip contains $3n^2/4$ leaves and each outer strip contains $n^2/8$ leaves. Similarly partition the layout into three horizontal strips H_0, H_1 and H_2 so that the middle strip contains $3n^2/4$ leaves and each outer strip contains $n^2/8$ leaves. For example, see Figure 8-1.

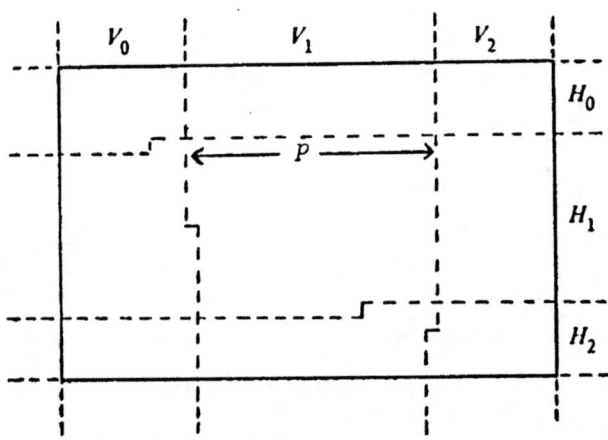

Figure 8-1: *Partitioning of the layout for* $M_{2,n}$.

Let p denote the length of the longest side of the center block formed by the intersection of V_1 and H_1. Without loss of generality, we assume that the longest side is horizontal. In what follows, we will show that $p \geq (\alpha^{1/2} n log n)/8$.

Since each of the regions $V_0 \cap H_1$ and $V_2 \cap H_1$ can contain at most $n^2/8$

leaves, it is clear that $V_j \cap H_j$ contains at least $n^2/2$ leaves. Consider the $n^{3/2}$ subgraphs of $M_{2,n}$ produced by eliminating the top $(3logn)/4$ levels of the row and column binary trees of $M_{2,n}$. Each of these subgraphs is isomorphic to $M_{2,n^{1/4}}$. By the pigeonhole principle, at least $1/2$ of these subgraphs have at least one leaf in $V_j \cap H_j$. If $p < (\alpha^{1/2}nlogn)/8$ (otherwise we are done), then at most $4p < (\alpha^{1/2}nlogn)/2$ edges can cross the boundary of $V_j \cap H_j$. Thus at most $(\alpha^{1/2}nlogn)/2$ of the subproblems which have at least one leaf in $V_j \cap H_j$ can have some node or part of an edge outside $V_j \cap H_j$. This means that at least $(n^{3/2} - \alpha^{1/2}nlogn)/2$ copies of $M_{2,n^{1/4}}$ are wholly contained in $V_j \cap H_j$. Applying the inductive hypothesis, we conclude that $V_j \cap H_j$ contains at least

$$(n^{3/2} - \alpha^{1/2}nlogn) \, W(n^{1/4})/2 \; \geq \; (\alpha n^2 log^2 n - \alpha^{3/2}n^{3/2}log^3 n)/32$$

$$\geq \; (\alpha n^2 log^2 n)/64 \quad \text{wire.}$$

(The last inequality follows trivially from (*).) Thus $V_j \cap H_j$ has at least $(\alpha n^2 log^2 n)/64$ area and $p \geq (\alpha^{1/2}nlogn)/8$, as claimed.

We next use the layout for $M_{2,n}$ to construct a drawing for the complete graph on n^2 nodes (namely, the n^2 leaves of $M_{2,n}$). No matter how the edges of the complete graph are drawn in the plane (e.g., they may cross or overlap), it is clear from Figure 8-1 that the sum of the lengths of all the edges (as measured in Euclidean space) is at least $n^4 p/64 \geq (\alpha^{1/2}n^5 logn)/2^9$. This is due to the fact that $n^4/64$ edges pass from region V_0 to region V_2 and that these regions are separated by a distance p.

Let L_i denote the sum of the lengths of the edges in the ith levels of the binary trees of $M_{2,n}$. Since every level i edge is traced over at most $n^3 2^{-i}$ times in the drawing of the complete graph, we can conclude that

$$\sum_{i=1}^{logn} L_i 2^{-i} n^3 \; \geq \; (\alpha^{1/2}n^5 logn)/2^9$$

and thus that

$$\sum_{i=1}^{logn} L_i 2^{-i} \geq (\alpha^{1/2} n^2 logn)/2^9 .$$

In particular, this means that

$$L_i \geq (\alpha^{1/2} n^2 logn 2^i)/(2^9 \beta i^2)$$

for some $i < logn$. (Recall that $\beta = \sum_{j=1}^{\infty} j^{-2}$.)   Otherwise,

$$L_i < (\alpha^{1/2} n^2 logn 2^i)/(2^9 \beta i^2)$$

for $1 \leq i \leq logn$ and thus

$$\sum_{i=1}^{logn} L_i 2^{-i} < \sum_{i=1}^{logn} (\alpha^{1/2} n^2 logn)/(2^9 \beta i^2)$$

$$\leq (\alpha^{1/2} n^2 logn)/2^9, \quad \text{a contradiction.}$$

Using the straightforward relation

$$W(n) \geq 2^{2i} W(n 2^{-i}) + L_i$$

where i has been chosen so that

$$L_i \geq (\alpha^{1/2} n^2 logn 2^i)/(2^9 \beta i^2),$$

we can conclude that

$$W(n) \geq 2^{2i} \alpha (n 2^{-i})^2 (logn - i)^2 + (\alpha^{1/2} n^2 logn 2^i)/(2^9 \beta i^2)$$

$$\geq \alpha n^2 log^2 n - 2\alpha i n^2 logn + (\alpha^{1/2} n^2 logn 2^i)/(2^9 \beta i^2)$$

$$\geq \alpha n^2 log^2 n .$$

(The last inequality follows trivially from (**).) Thus $W(n) \geq \Omega(n^2 log^2 n)$ for all n. $\square$

Theorem 8-2: *Any layout of the N-node 2-dimensional mesh of trees contains a wire of length at least $\Omega(N^{1/2} logN/loglogN)$.*

Proof: It is sufficient to show that any layout for $M_{2,n}$ contains a wire of length at least $\Omega(nlogn/loglogn)$. Assume for the purposes of contradiction that this is not the case and consider a layout of $M_{2,n}$ for which the longest wire has length

$q \ll O(nlogn/loglogn)$.   Using arguments similar to those used to prove Theorem 5-2, we first show that (without loss of generality) the area of such a layout is at most $O(q^2log^2n) \ll O(n^2log^4n)$.

Since every pair of nodes of $M_{2,n}$ is linked by a path of length at most $4logn$, all of the nodes in the layout are contained in a $4qlogn \times 4qlogn$ square. At most $16qlogn$ wires may leave and re-enter the square at various points along its boundary. Without increasing the lengths of any of these wires, it is possible to rewire the segments outside the square using at most $O(q^2log^2n)$ additional area. Thus, the resulting layout for $M_{2,n}$ will have maximum edge length q and area at most $O(q^2log^2n)$.

The proof is completed by observing that any layout of $M_{2,n}$ with area less than $O(n^2log^4n)$ must have a wire of length at least $\Omega(nlogn/loglogn)$. From the proof of Theorem 8-1. we know that $\sum_{i=1}^{logn} L_i 2^{-i} \geq (\alpha^{1/2}n^2logn)/2^9$ Thus either

1) there is an $i \leq 4loglogn$ such that $L_i \geq (\alpha^{1/2}n^2logn2^i)/(2^{12}loglogn)$,

or

2) there is an $i > 4loglogn$ such that $L_i \geq (\alpha^{1/2}n^2logn2^i)/(2^{10}\beta i^2)$

where, as before, the constant $\beta = \sum_{j=1}^{\infty} j^{-2}$. Otherwise,

$$\sum_{i=1}^{logn} L_i 2^{-i} = \sum_{i=1}^{4loglogn} L_i 2^{-i} + \sum_{i=4loglogn+1}^{logn} L_i 2^{-i}$$
$$< (\alpha^{1/2}n^2logn)/2^{10} + [(\alpha^{1/2}n^2logn)/2^{10}\beta] \sum_{i=4loglogn+1}^{logn} i^{-2}$$
$$\leq (\alpha^{1/2}n^2logn)/2^9 , \quad \text{a contradiction.}$$

The second condition cannot possibly be true, however. If it were, the area of the layout would be at least

$$L_i \geq \Omega(n^2logn/i^2)$$
$$\geq \Omega(n^2log^5n/(loglogn)^2)$$
$$> \Omega(n^2log^4n) , \quad \text{a contradiction.}$$

Thus the first condition must be true and there is an i such that $L_i \geq \Omega(n^2 log n 2^i / log log n)$. Since there are $n 2^{i+1}$ type i edges in $M_{2,n}$, we can conclude that at least one of them has length at least $\Omega(n log n / log log n)$. $\square$

8.2   Lower Bounds for the Tree of Meshes

Using the results of the previous section, it is easy to demonstrate the existence of planar graphs which *cannot* be laid out in linear area and which *must* have long wires. In particular, we can conclude the following.

Theorem 8-3: *The wire area of the N-node tree of meshes is at least $\Omega(N log N)$.*

Proof: As we showed in Section 6.3.3b, the N-node 2-dimensional mesh of trees can be embedded in an $O(N log N)$-node tree of meshes. By Theorem 8-1, we can thus conclude that the wire area of the $N log N$-node tree of meshes is at least $\Omega(N log^2 N)$. Equivalently, the wire area of the N-node tree of meshes is at least $\Omega(N log N)$. $\square$

Theorem 8-4: *Any layout of the N-node augmented tree of meshes has a wire of length at least $\Omega(N^{1/2} / log^{1/2} N)$.*

Proof: In the proof of Theorem 8-1, we showed that any layout of $M_{2,n}$ has two leaves which are spaced at least $\Omega(n log n)$ distance apart. Since (as we showed in Section 6.3.3b) $M_{2,n}$ can be embedded in T_{2n} so that the leaves of $M_{2,n}$ are embedded in the leaves of T_{2n}, we can observe that any layout of T_{2n} also has two leaves which are spaced at least $\Omega(n log n)$ distance apart. Since every pair of leaves in T_{2n} are linked by a path of length at most $O(log n)$ in T_{2n}', we can conclude that some edge of T_{2n}' has length at least $\Omega(n) = \Omega(N^{1/2} / log^{1/2} N)$. $\square$

Had we so desired, we could have proved both results directly, using arguments identical to the ones used to prove Theorem 8-1.

8.3   Lower Bounds for a Restricted Class of Binary Tree Layouts

In [15], Brent and Kung considered layouts of N-node complete binary trees for which every leaf is located on the boundary of some convex region. In particular, they showed that the wire area of any such layout is at least $\Omega(N\log N)$. Recently, Paterson, Ruzzo and Snyder [67] extended this result by showing that any such layout with area A must have some edge of length $\Omega(N/\log(A/N))$. In particular, this means that if $A = O(N\log N)$, then there must be some edge of length $\Omega(N/\log\log N)$ but that if $A = \Theta(N^{1+\epsilon})$ for some $\epsilon > 0$, then there must only be an edge of length $\Omega(N/\log N)$. In what follows, we show how to use the techniques developed in this chapter to give short proofs of these facts.

Theorem 8-5 (Brent and Kung [15]):   *Any layout of the N-node complete binary tree in which every leaf is on the boundary of some convex region requires $\Omega(N\log N)$ area.*

Proof: Given any such layout, we first use the methods of Section 8.1 to construct a layout of the complete graph on the $n = \Theta(N)$ leaves of the tree. Since the leaves are on the boundary of some convex region, it is easily shown that the layout of K_n uses at least $\Omega(n^3)$ wire.

Let L_i denote the sum of the lengths of the edges in the ith level of the tree. As each ith level edge is traced over at most $n^2 2^{-i}$ times, we know that

$$\sum_{i=1}^{\log n} n^2 2^{-i} L_i \leq \Omega(n^3)$$

and thus that $\sum_{i=1}^{\log n} L_i 2^{-i} \geq \Omega(n)$. Using arguments similar to those in the proof of Theorem 8-1, we can conclude that $L_i \geq \Omega(n 2^i / i^2)$ for at least one value of i.

Letting $W(n)$ denote the wire area of the binary tree layout, we can see that

$$W(n) \geq 2^i W(n2^{-i}) + \Omega(n2^i/i^2).$$

Solving the recurrence, we find that $W(n) \geq \Omega(n \log n) = \Omega(N \log N)$. $\square$

Theorem 8-6 (Paterson, Ruzzo and Snyder [67]): *Any A-area layout of the N-node complete binary tree in which every leaf is on the boundary of a convex region has some edge of length* $\Omega(N/\log(A/N))$.

Proof: The proof follows that of the preceding theorem until it is concluded that $\sum_{i=1}^{\log n} L_i 2^{-i} \geq \Omega(n)$. Using methods similar to those used to prove Theorem 8-2, we can then observe that one of the following conditions must be satisfied:

1) there is an $i \leq 2\log(A/n)$ such that $L_i \geq \Omega(n2^i/\log(A/n))$, or

2) there is an $i \geq 2\log(A/n)$ such that $L_i \geq \Omega(n2^i/i^2)$.

The second condition cannot possibly hold since, if it did, the layout area would be at least $L_i \geq \Omega(n2^i/i^2)$ which, for $i \geq 2\log(A/n)$, means that

$$A \geq \Omega(A^2/n\log^2(A/n))$$

$$> \Omega(A), \quad \text{a contradiction.}$$

Thus the first condition holds and we can conclude that there is an i such that $L_i \geq \Omega(n2^i/\log(A/n))$. As there are only 2^{i+1} edges in the ith level, at least one of them must have length at least $\Omega(n/\log(A/n)) = \Omega(N/\log(A/N))$. $\square$

ADDENDUM

Much has been accomplished since the first draft of this book was completed. In fact, so much has been recently discovered that it would be possible to write several additional books on the subject. In what follows, we highlight some of the important recent developments and give pointers to relevant research.

General Layout Strategies. Bhatt and Leighton recently developed a framework for solving general graph layout problems. The new framework is based on the notion of a *bifurcator* and provides simple solutions to a number of layout-related problems that were previously thought to be difficult. The work represents a substantial improvement over previous separator-based techniques and generalizes several of the results contained in Part II of this book. Descriptions of the new work can be found in [8, 49]. Related material can be found in [3, 9, 10, 25, 36, 37, 67, 84, 100].

Crossing Number Arguments. Garey and Johnson recently showed that the problem of determining the minimum crossing number of a graph is NP-complete [29]. Obtaining an approximation algorithm for the crossing number problem remains an important open problem, although Leighton [49] reduced the task to that of finding an approximation algorithm for the bisection width problem (another important open problem). Lastly, Aggarwal used crossing number arguments like those in Chapter 7 to prove time lower bounds for certain non-unit-delay models of parallel computation [2].

Algorithms for the Mesh of Trees. A number of researchers independently discovered the mesh of trees and/or developed parallel algorithms based on its structure. Nath, Maheshwari and Bhatt used the 2-dimensional mesh of trees (which they call the *orthogonal trees network*) for sorting, discrete Fourier transform, minimum spanning tree and connected components (as well as other) problems [65]. Cappello and Steiglitz used a variant of the 2-dimensional mesh of trees (which they call the *orthogonal forests*) for integer multiplication [19]. Preparata and Vuillemin used the 3-dimensional mesh of trees for matrix multiplication and were the first to discover the algorithm we described in Chapter 6 [75]. Lastly, Leighton used the shuffle-tree graph to find new algorithms for permuting lists [50].

Layouts With Fixed Placement. The layout problem studied in this book consists of two phases: placement of the nodes, and routing of the wires. Recently, Karp, Leighton, Rivest, Thompson, Vazirani and Vazirani found approximation algorithms for the routing portion of this problem (i.e., given a fixed placement) [36]. The results seem to indicate that the placement phase is the most difficult part of the layout problem.

Channel Routing. Channel routing plays a key role in most practical layout schemes and has recently been studied by a number of theoretical researchers. Rivest, Baratz and Miller developed an approximation algorithm for knock-knee routing in [77]. Leighton proved the optimality (by example) of this algorithm in [48]. Preparata and Lipski [73] studied the 3-layer version of the knock-knee problem and discovered an optimal routing algorithm in the case of 2-point nets. This work is currently being extended to handle multipoint nets by Brown and Preparata [17]. Baker, Bhatt and Leighton discovered an approximation algorithm for Manhattan routing multipoint nets [4]. (This problem had been shown previously to be NP-complete by Szymanski in [91].) Lastly, Pinter (in one case working with Leiserson) discovered optimal solutions to a number of river routing problems [57, 68, 69, 71]. Reports on additional channel routing work can be found in [13, 14, 18, 24, 44, 45, 70, 76, 78, 96, 102].

Three-Dimensional Layouts. Leighton and Rosenberg extended a number of the results in this book to three dimensions [54]. Although three-dimensional circuits are more powerful than two-dimensional circuits, there are some surprising limitations to their functionality. Other work on three-dimensional layouts can be found in [72, 81].

Construction of Networks Around Faults. Throughout this book, we have considered layout problems for which processors can be arbitrarily located throughout the grid. A number of researchers have since considered models for which faults in the grid severely restrict the location of processors. Reports on this research can be found in [7, 22, 28, 30, 33, 41, 53, 80, 82].

REFERENCES

[1] H. Abelson and P. Andreae, "Information transfer and area-time tradeoffs for VLSI multiplication," *Communications of the ACM*, Vol. 23, 1980, pp. 20-23.

[2] A. Aggarwal, "Period-time tradeoffs for VLSI models with delay," *Proceedings of the 24th Annual IEEE Symposium on Foundations of Computer Science*, November 1983, to appear.

[3] R. Aleliunas and A. L. Rosenberg, "On embedding rectangular grids in square grids," *IEEE Transactions on Computers*, Vol. C-31, 1980, pp. 907-913.

[4] B. S. Baker, S. Bhatt and F. T. Leighton, "An approximation algorithm for Manhattan routing," *Proceedings of the 15th Annual ACM Symposium on Theory of Computing*, April 1983, pp. 477-486.

[5] G. M. Baudet, "On the area required by VLSI circuits," *Proceedings of the CMU Conference on VLSI Systems and Computations*, edited by H. T. Kung, B. Sproull and G. Steele, Computer Science Press, Rockville Maryland, October 1981, pp. 100-107.

[6] C. M. Bender and S. A. Orszag, *Advanced Mathematical Methods for Scientists and Engineers*, McGraw-Hill Book Company, New York, 1978.

[7] F. Berman, F. T. Leighton and L. Snyder, "Optimal tile salvage," MIT-VLSI Technical Memo #82-119.

[8] S. N. Bhatt and F. T. Leighton, "A framework for solving VLSI graph layout problems," *Journal of Computer and System Sciences*, to appear.

[9] S. N. Bhatt and C. E. Leiserson, "Minimizing the longest edge in a VLSI layout," MIT-VLSI Technical Memo #82-86, 1982.

[10] S. N. Bhatt and C. E. Leiserson, "How to assemble tree machines," *Proceedings of the 14th Annual ACM Symposium on Theory of Computing*, May 1982, pp. 77-84.

[11]   N. L. Biggs, E. K. Lloyd and R. J. Wilson, *Graph Theory 1736-1936*, Clarendon Press, Oxford, 1976.

[12]   G. Bilardi, M. Pracchi and F. P. Preparata, "A critique and appraisal of VLSI models of computation," *Proceedings of the CMU Conference on VLSI Systems and Computations*, edited by H. T. Kung, B. Sproull and G. Steele, Computer Science Press, Rockville Maryland, October 1981, pp. 81-88.

[13]   T. Bolognesi, *A Channel Routing Algorithm Bounding Channel Width and Maximum Wire Length*, M. S. Thesis, University of Illinois, 1982.

[14]   T. Bolognesi and D. Brown, "A channel routing algorithm with bounded wire length," unpublished manuscript, University of Illinois, 1982.

[15]   R. P. Brent and H. T. Kung, "On the area of binary tree layouts," *Information Processing Letters*, No. 11, 1980, pp. 44-46.

[16]   R. P. Brent and H. T. Kung, "The chip complexity of binary arithmetic," *Proceedings of the 12th Annual ACM Symposium on Theory of Computing*, April 1980, pp. 190-200.

[17]   D. Brown and F. P. Preparata, personal communication, University of Illinois, 1982.

[18]   D. Brown and R. L. Rivest, "New lower bounds on channel width," *Proceedings of the CMU Conference on VLSI Systems and Computations*, edited by H. T. Kung, B. Sproull and G. Steele, Computer Science Press, Rockville Maryland, October 1981, pp. 178-185.

[19]   P. R. Cappello and K. Steiglitz, "Area-efficient VLSI structures for multiplying at clock rate," Technical Report #289, Department of EECS, Princeton University, September 1981.

[20]   B. Chazelle and L. Monier, "A model of computation for VLSI with related complexity results," *Proceedings of the 13th Annual ACM Symposium on Theory of Computing*, May 1981, pp. 318-325.

[21]   B. Chazelle and L. Monier, "Optimality in VLSI," *VLSI 81: Very Large Scale Integration*, edited by J. P. Gray, Academic Press, London, August 1981, pp. 269-278.

[22] F. R. K. Chung, F. T. Leighton and A. L. Rosenberg, "Diogenes: a methodology for designing fault-tolerant VLSI processor arrays," *Proceedings of the 1983 International Conference on Fault-Tolerant Computing*, Milan, Italy, to appear.

[23] P. Diaconis, R. L. Graham and W. M. Cantor, "The mathematics of perfect shuffles," unpublished manuscript, Bell Labs, 1981.

[24] D. Dolev, K. Karplus, A. Siegel, A. Strong and J. Ullman, "Optimal wiring between rectangles," *Proceedings of the 13th Annual ACM Symposium on Theory of Computing*, May 1981, pp. 312-317.

[25] D. Dolev, F. T. Leighton and H. Trickey, "Planar embeddings of planar graphs," MIT-LCS Technical Memo #237, 1983.

[26] M. J. Fischer and M. S. Paterson, "Optimal tree layout," *Proceedings of the 12th Annual ACM Symposium on Theory of Computing*, April 1980, pp. 177-189.

[27] R. W. Floyd and J. D. Ullman, "The compilation of regular expressions into integrated circuits," *Proceedings of the 21st Annual IEEE Symposium on Foundations of Computer Science*, October 1980, pp. 260-269.

[28] D. Fussel and P. Varman, "Fault-tolerant wafer-scale architectures for VLSI," *Proceedings of the 9th Annual IEEE/ACM Symposium on Computer Architecture*, April 1982, pp. 190-198.

[29] M. Garey and D. Johnson, "Crossing number is NP-complete," unpublished manuscript, Bell Labs, December 1981.

[30] J. Greene and A. El-Gamal, "Area and delay penalties for restructurable VLSI arrays," unpublished manuscript, Stanford University, May 1982.

[31] L. J. Guibas and A. M. Odlyzko, "Periods in strings," *Journal of Combinatorial Theory (Series A)*, Vol. 30, No. 1, January 1981, pp. 19-42.

[32] L. J. Guibas and A. M. Odlyzko, "String overlaps, pattern matching and nontransitive games," *Journal of Combinatorial Theory (Section A)*, Vol. 30, 1981, pp. 183-208.

[33] K. Hedlund, personal communication, University of North Carolina, November 1982.

[34]   D. Hoey and C. E. Leiserson, "A layout for the shuffle-exchange network,"
        Proceedings of the 1980 IEEE Conference on Parallel Processing, August
        1980.

[35]   H. Jia-Wei and A. L. Rosenberg, "Graphs that are similar to binary trees,"
        Proceedings of the 13th Annual ACM Symposium on Theory of Computing,
        May 1981, pp. 334-341.

[36]   R. M. Karp, F. T. Leighton, R. L. Rivest, C. D. Thompson, U. Vazirani and
        V. Vazirani, "Global wire routing in two-dimensional arrays," *Proceedings
        of the 24th Annual IEEE Symposium on Foundations of Computer Science,*
        November 1983, to appear.

[37]   Z. M. Kedem, "Optimal allocation of computational resources in VLSI,"
        *Proceedings of the 23rd Annual IEEE Symposium on Foundations of
        Computer Science,* November 1982, pp. 379-385.

[38]   Z. M. Kedem and A. Zorat, "On relations between input and
        communication/computation in VLSI," *Proceedings of the 22nd Annual
        IEEE Symposium on Foundations of Computer Science,* October 1981, pp.
        37-44.

[39]   D. J. Kleitman, "The crossing number of $K_{5,n}$," *Journal of Combinatorial
        Theory,* Vol. 9, No. 4, December 1970, pp. 315-323.

[40]   D. J. Kleitman, F. T. Leighton, M. Lepley and G. L. Miller, "New layouts
        for the shuffle-exchange graph," *Proceedings of the 13th Annual ACM
        Symposium on Theory of Computing,* May 1981, pp. 278-292. (To appear as
        "An asymptotically optimal layout for the shuffle-exchange graph" in the
        Journal of Computer and Systems Science.)

[41]   I. Koren, "A reconfigurable and fault-tolerant VLSI multiprocessor array,"
        *Proceedings of the 8th Annual IEEE/ACM Symposium on Computer
        Architecture,* May 1981, pp. 425-431.

[42]   H. T. Kung and C. E. Leiserson, "Algorithms for VLSI processor arrays,"
        Symposium on Sparse Matrix Computations, Knoxville Tennessee,
        November 1978.

[43]   T. Lang, "Interconnection between processing and memory modules using
        the shuffle-exchange network," *IEEE Transactions on Computers,* Vol. C-
        25, January 1976, pp. 55-66.

[44]   A. S. LaPaugh, *Algorithms for Integrated Circuit Layout: an Analytic Approach*, Ph. D. Thesis, Department of Electrical Engineering and Computer Science, MIT, 1980.

[45]   A. S. LaPaugh, "A polynomial time algorithm for optimal routing around a rectangle, " *Proceedings of the 21st Annual IEEE Symposium on Foundations of Computer Science*, October 1980, pp. 282-293.

[46]   F. T. Leighton, *Layouts for the Shuffle-Exchange Graph and Lower Bound Techniques for VLSI*, Pd. D. Thesis, Mathematics Department, Massachusetts Institute of Technology, Cambridge Massachusetts, September 1981.

[47]   F. T. Leighton, "New lower bound techniques for VLSI," *Proceedings of the 22nd Annual IEEE Symposium on Foundations of Computer Science*, October 1981, pp. 1-12.   (To appear in *Math Systems Theory*.)

[48]   F. T. Leighton, "New lower bounds for channel routing," MIT-VLSI Technical Memo #82-71, 1982.

[49]   F. T. Leighton, "A layout strategy for VLSI which is provably good," *Proceedings of the 14th Annual ACM Symposium on Theory of Computing*, May 1982, pp. 85-98.

[50]   F. T. Leighton, "Parallel computation using meshes of trees," *Proceedings of the 1983 Workshop on Graphtheoretic Concepts in Computer Science*, Osnabruck, West Germany, to appear.

[51]   F. T. Leighton, M. Lepley and G. L. Miller, "Layouts for the shuffle-exchange graph based on the complex plane diagram," *SIAM Journal of Algebraic and Discrete Methods*, to appear.

[52]   F. T. Leighton and G. L. Miller, "Optimal layouts for small shuffle-exchange graphs," *VLSI 81 - Very Large Scale Integration*, edited by John P. Gray, Academic Press, London, August 1981, pp. 289-299.

[53]   F. T. Leighton and C. E. Leiserson, "Wafer-scale integration of systolic arrays," *Proceedings of the 23rd Annual IEEE Symposium on Foundations of Computer Science*, November 1982, pp. 297-310.

[54]   F. T. Leighton and A. Rosenberg, "Three-dimensional circuit layouts," *Proceedings of the 1983 IEEE International Conference on Computer Design,* Rye New York, to appear.

[55]   C. E. Leiserson, "Area-efficient graph layouts (for VLSI)," *Proceedings of the 21st Annual IEEE Symposium on Foundations of Computer Science,* October 1980, pp. 270-281.

[56]   C. E. Leiserson, *Area Efficient VLSI Computation,* MIT Press, Cambridge Massachusetts, 1983.

[57]   C. E. Leiserson and R. Y. Pinter, "Optimal placement for river routing," *Proceedings of the CMU Conference on VLSI Systems and Computations,* edited by H. T. Kung, B. Sproull and G. Steele, Computer Science Press, Rockville Maryland, October 1981, pp. 126-142.

[58]   T. Lengauer and K. Mehlhorn, "On the complexity of VLSI computations," *Proceedings of the CMU Conference on VLSI Systems and Computations,* edited by H. T. Kung, B. Sproull and G. Steele, Computer Science Press, Rockville Maryland, October 1981, pp. 89-99.

[59]   R. J. Lipton and R. Sedgewick, "Lower bounds for VLSI," *Proceedings of the 13th Annual ACM Symposium on Theory of Computing,* May 1981, pp. 300-307.

[60]   R. J. Lipton and R. E. Tarjan, "A separator theorem for planar graphs," *A Conference on Theoretical Computer Science,* University of Waterloo, August 1977.

[61]   C. Mead and L. Conway, *Introduction to VLSI Systems,* Addison-Wesley Publishing Company, Reading Massachusetts, October 1980.

[62]   C. Mead and M. Rem, "Cost performance of VLSI computing structures," *IEEE Journal of Solid State Circuits,* Vol. SC-14, No. 2, April 1979, pp. 455-462.

[63]   D. E. Muller and F. P. Preparata, "Bounds to complexities of networks for sorting and for switching," *Journal of the ACM,* Vol. 22, No. 2, April 1975, pp. 195-201.

[64]   D. Nassimi and S. Sahni, "A self-routing Benes network and parallel permutation alogorithms," University of Minnesota Technical Report 79-13, May 1979.

[65]   D. Nath, S. N. Maheshwari and P. C. P. Bhatt, "Efficient VLSI networks for parallel processing based on orthogonal trees," unpublished manuscript, 1981.

[66]   D. S. Parker, "Notes on shuffle/exchange-type switching networks," *IEEE Transactions on Computers*, Vol. C-29, No. 3, March 1980, pp. 213-222.

[67]   M. S. Paterson, W. L. Ruzzo and L. Snyder, "Bounds on minimax edge length for complete binary trees," *Proceedings of the 13th Annual ACM Symposium on Theory of Computing*, May 1981, pp. 293-299.

[68]   R. Y. Pinter, "Optimal routing in rectilinear channels," *Proceedings of the CMU Conference on VLSI Systems and Computations*, edited by H. T. Kung, B. Sproull and G. Steele, Computer Science Press, Rockville Maryland, October 1981, pp. 160-177.

[69]   R. Y. Pinter, "On routing 2-point nets across a channel," *Proceedings of the 19th Annual ACM/IEEE Design Automation Conference*, June 1982, pp. 894-902.

[70]   R. Y. Pinter, "Optimal layer assignment for interconnect," *Proceedings of the IEEE Conference on Circuits and Computers*, October 1982, pp. 398-401.

[71]   R. Y. Pinter, *The Impact of Layer Assignment Methods on Layout Algorithms for Integrated Circuits*, Ph. D. Thesis, Department of Electrical Engineering and Computer Science, MIT, December 1982.

[72]   F. P. Preparata, "Optimal three-dimensional VLSI layouts," *Math Systems Theory*, to appear.

[73]   F. P. Preparata and W. Lipski, "Three layers are enough," *Proceedings of the 23rd Annual IEEE Conference on Foundations of Computer Science*, November 1982.

[74]   F. P. Preparata and J. E. Vuillemin, "The cube-connected-cycles: a versatile network for parallel computation," *Proceedings of the 20th Annual IEEE Symposium on Foundations of Computer Science*, October 1979, pp. 140-147.

[75] F. P. Preparata and J. E. Vuillemin, "Area-time optimal VLSI networks for matrix multiplication," *Proceedings of the 14th Princeton Conference on Information Science and Systems,* 1980.

[76] R. L. Rivest, "The PI (placement and interconnect) system," *Proceedings of the 19th Annual ACM/IEEE Design Automation Conference,* June 1982, pp. 475-481.

[77] R. L. Rivest, A. Baratz and G. L. Miller, "Provably good channel routing algorithms," *Proceedings of the CMU Conference on VLSI Systems and Computations,* edited by H. T. Kung, B. Sproull and G. Steele, Computer Science Press, Rockville Maryland, October 1981, pp. 153-159.

[78] R. L. Rivest and C. M. Fidiccia, "A greedy channel router," *Proceedings of the 19th Annual ACM/IEEE Design Automation Conference,* June 1982, pp. 418-424.

[79] A. L. Rosenberg, "On embedding graphs in grids," IBM Watson Research Center Technical Report RC7559 (#2668), 1979.

[80] A. L. Rosenberg, "Routing with permuters: toward reconfigurable and fault-tolerant networks," Duke University Technical Report CS-1981-13, 1981.

[81] A. L. Rosenberg, "Three-dimensional integrated circuitry," *Proceedings of the CMU Conference on VLSI Systems and Computations,* edited by H. T. Kung, B. Sproull and G. Steele, Computer Science Press, Rockville Maryland, October 1981, pp. 69-80.

[82] A. L. Rosenberg, "The Diogenes approach to testable fault-tolerant networks of processors," Duke University Technical Report CS-1982-6.1, May 1982.

[83] A. L. Rosenberg, "On designing fault-tolerant arrays of processors," Duke University Technical Report CS-1982-14.

[84] W. L. Ruzzo and L. Snyder, "Minimum edge length planar embeddings of trees," *Proceedings of the CMU Conference on VLSI Systems and Computations,* edited by H. T. Kung, B. Sproull and G. Steele, Computer Science Press, Rockville Maryland, October 1981, pp. 119-123.

[85] J. E. Savage, "Area-time tradeoffs for matrix multiplication and related problems in VLSI models," *Proceedings of the 17th Annual Allerton Conference on Communications, Control and Computing,* October 1979, pp. 670-676.

[86] J. E. Savage, "Planar circuit complexity and the performance of VLSI algorithms," *Proceedings of the CMU Conference on VLSI Systems and Computations,* edited by H. T. Kung, B. Sproull and G. Steele, Computer Science Press, Rockville Maryland, October 1981, pp. 61-68.

[87] J. T. Schwartz, "Ultracomputers," *ACM Transactions on Programming Languages and Systems,* Vol. 2, No. 4, October 1980, pp. 484-521.

[88] L. Snyder, "Overview of the CHiP computer," *VLSI 81 - Very Large Scale Integration,* edited by J. Gray, Academic Press, London, August 1981, pp. 237-246.

[89] D. Steinberg and M. Rodeh, "A layout for the shuffle-exchange network with $O(N^2/\log^{3/2}N)$ area," *IEEE Transactions on Computers,* to appear.

[90] H. S. Stone, "Parallel processing with the perfect shuffle," *IEEE Transactions on Computers,* Vol. C-20, No. 2, February 1971, pp. 153-161.

[91] T. Szymanski, "Dogleg channel routing is NP-complete," unpublished manuscript, Bell Labs, 1982.

[92] C. D. Thompson, "Area-time complexity for VLSI," *Proceedings of the 11th Annual ACM Symposium on Theory of Computing,* May 1979, pp. 81-88.

[93] C. D. Thompson, *A Complexity Theory for VLSI,* Ph.D. Thesis, Department of Computer Science, Carnegie-Mellon University, 1980.

[94] C. D. Thompson, "The VLSI complexity of sorting," *Proceedings of the CMU Conference on VLSI Systems and Computations,* edited by H. T. Kung, B. Sproull and G. Steele, Computer Science Press, Rockville Maryland, October 1981, pp. 108-118.

[95] C. D. Thompson and H. T. Kung, "Sorting on a mesh-connected parallel computer," *Communications of the ACM,* Vol. 20, 1977, pp. 263-271.

[96]   M. Tompa, "An optimal solution to a wire-routing problem," *Proceedings of the 12th Annual ACM Symposium on Theory of Computing*, April 1980, pp. 161-176.

[97]   H. W. Trickey, "Good layouts for pattern recognizers," *IEEE Transactions on Computers*, Vol. C-31, pp. 514-520.

[98]   L. G. Valiant, "Universality considerations in VLSI circuits," *IEEE Transactions on Computers*, Vol. V-30, No. 2, February 1981, pp. 135-140.

[99]   J. E. Vuillemin, "A combinatorial limit to the computing power of VLSI circuits," *Proceedings of the 21st Annual Symposium on Foundations of Computer Science*, October 1980, pp. 294-300.

[100] D. Wise, "Efficient layouts for FFT and Banyan networks," *Proceedings of the CMU Conference on VLSI Systems and Computations*, edited by H. T. Kung, B. Sproull and G. Steele, Computer Science Press, Rockville Maryland, October 1981, pp. 186-195.

[101] A. C. Yao, "The entropic limitations on VLSI Computations," *Proceedings of the 13th Annual ACM Symposium on Theory of Computing*, May 1981, pp. 308-311.

[102] T. Yoshimura and E. Kuh, "Efficient algorithms for channel routing," U. C. Berkeley Electronics Research Lab Memo M80/43, 1980.

INDEX OF DEFINITIONS